Module 2

Operations on Polynomials
Second Edition

Operations on Polynomials
Second Edition

Leon J. Ablon   Sherry Blackman   Helen B. Siner
The College of Staten Island, City University of New York

Anthony Giangrasso
La Guardia Community College, City University of New York

The Benjamin/Cummings Publishing Company, Inc.
Menlo Park, California • Reading, Massachusetts
London • Amsterdam • Don Mills, Ontario • Sydney

Sponsoring Editor: Susan A. Newman
Production Editor: Madeleine Dreyfack
Book Designer: Madeleine Dreyfack
Cover Designer: Judith Sager

TO

Fannie and Meyer Ablon
Michael Blackman
Mr. & Mrs. Adam Kowalski
Muriel and Matthew Brody

Library of Congress Cataloging in Publication Data
Main entry under title:

Operations on polyomials.

    (Series in mathematics modules ; module 2)
    1. Polynomials. I. Ablon, Leon J. II. Series.
QA39.2.S47 no. 2, 1980 [QA161.P59] 512.9s [512.9'42]
ISBN 0-8053-0132-1   80-24843

ABCDEFGHIJ-DO-83210

The Benjamin/Cummings Publishing Company, Inc.
2727 Sand Hill Road
Menlo Park, California 94025

Preface

Purpose of the Steps in Mathematics Modules

This book is one of the Steps in Mathematics Modules. The purpose of this series is to demystify mathematics and provide a foundation for the study of college algebra. Throughout our years of teaching we have become sensitive to the difficulties our students encounter in their mathematics courses. These modules incorporate the techniques we have evolved to overcome these problems.

Features

Development of Concepts: There are very few "rules" in these texts. Whenever possible, we encourage students to figure out what to do from the meaning of the symbols.

Writing Style: We have been as clear as possible in presenting each mathematical concept. We have found that students relate well to our use of everyday language. Our guiding principle has been to use language and syntax no more complex than is necessary to convey each concept. It is the kind of language that is used in the classroom, but is rarely found in mathematics textbooks.

Organization: Each module is divided into eight lessons. Each lesson is designed to be covered in one class period.

Examples and Exercises: We have used many "worked out" examples in each lesson to demonstrate the mathematics. Each lesson ends with two sets of *graded* exercises. The first set is followed immediately by handwritten solutions, so the student can see what a typical solution might look like. The second set, or "additional" exercise set, does not have solutions in the module. The answers to these exercises can be found in the Rationale, a supplement to these modules.

How to Use the Modules

Pacing: Each module is made up of eight lessons, and contains material sufficient for a 40 to 50-minute class period. The background of some students may, of course, dictate a slower or faster pace. On the average we have found that each module can be completed in 10 to 12 class meetings, including time for review and examination.

Planning a Syllabus for a Lecture Mode: The chart below may assist you in setting up a program for a term. It shows the number of modules that can be covered in 10, 12 and 14-week terms with classes that meet 3, 4 and 5 times a week.

Weeks per term	Meetings per week		
	3	4	5
10	3 modules	3 or 4 modules	4 or 5 modules
12	3 modules	4 modules	5 modules
14	4 modules	5 modules	6 or 7 modules

Self-Study or Lab Mode of Instruction: The modules can be used in math lab or for independent study. The preceding chart serves as a guide for students to gauge their progress.

Flexibility: The modular presentation of material allows for flexibility in teaching. In particular, the concept of changing the form of expressions (Modules 2 and 4) is separated from the concept of solving equations (Modules 3 and 5).

A chart of the modules interdependence can be found in the Rationale. We recommend that each module be studied from cover to cover. This gives students a feeling of completeness and insures that they are prepared for succeeding modules.

The Diagnostic Test included in the Rationale is designed to assist in determining an appropriate starting module for each student.

Improvements in This Edition

The first edition of these modules has been used successfully by us and many other math teachers in classrooms and labs for over eight years. As a result of our experiences and helpful feedback from other teachers, we have made several changes and refinements in this second edition.

Decimals: This new edition integrates a review of decimals into both the lessons and homework exercises.

Word Problems: Word problems have been carefully integrated throughout the modules in the examples and exercises.

Exercises: We have increased the number of homework exercises by 20%.

Geometry: The series now includes a review of the concepts of perimeter, area, formulas and the Pythagorean theorem.

Design and Appearance: The modules are typeset, a pleasant change from the camera-ready typewriter version of the first edition. The use of italic and boldface type enables the student to quickly pick out examples, important concepts and key terms.

The Steps in Mathematics Modules Series

The first five steps in mathematics are the equivalent of an elementary algebra course: They are:

Module 1—Operations on Numbers

Module 2—Operations on Polynomials

Module 3—Linear Equations and Lines

Module 4—Factoring and Operations on Algebraic Fractions

Module 5—Quadratic Equations and Curves

Program Rationale and Tests—an instructor's supplement which contains a brief description of the content of each lesson, answers to the Additional Exercises, diagnostic placement examinations and three mastery tests for each of the five modules.

Additional modules in the **Steps in Mathematics Series** cover topics in intermediate algebra and other specialized topics.

Module 1a—Practical Mathematics

Module 2a—Practical Mathematics

Module 6 —Basic Trigonometry, second edition

Module 7 —Trigonometry with Applications

Module 8 —Exponents and Logarithms

Module 9 —Advanced Algebraic Techniques

Module 10—Functions and Word Problems

Module 11—Graphing Functions

Module M —Medical Dosage Calculations, second edition

Module SI—Metric System

Acknowledgements

We wish to thank all the people who contributed to the revision of Modules 1–5 with special thanks to our reviewers:

Dr. George Bergman, University of California, Berkeley
Dr. Una Bray, Marymount Manhattan College
Professor Ruth Dorsett, Atlanta Junior College
Dr. Susan Lawrence, New York University
Dr. Betty Philips, Michigan State University
Professor Michael Shaughnessy, Oregon State University
Professor Billie J. Stacey, Sinclair Community College

Our appreciation, also, to Susan Newman, sponsoring editor, and Madeleine Dreyfack, production editor and designer. The special efforts they made to translate our ideas into a finished work are greatly appreciated.

Leon J. Ablon
Sherry Blackman
Anthony P. Giangrasso
Helen B. Siner
October, 1980

To the Student

Read this text. It was written with the help of our students so students could read it. Our students tell us that the best way to learn this material is:

1. Read the lesson.
2. Work out each EXAMPLE yourself.
3. Try the first set of EXERCISES at the end of the lesson and compare your answers with ours.
4. Do the ADDITIONAL EXERCISES for more practice.

We wish you success, and we would like to hear your reactions to the text and your suggestions for future editions.

Contents

Review of Operations on Signed Numbers

A temperature of ten below zero is negative ten, and we write it "−10." A temperature of ten above zero is positive ten and we write it "+10." It is useful to arrange numbers on a number line as in Figure 1.

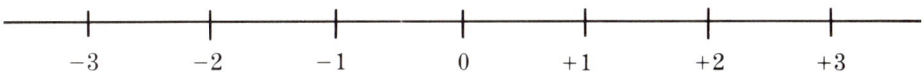

Figure 1

Figure 1 shows only a part of a number line, a part that fits on the page. A number line actually has no beginning or end.

Although in Figure 1 we labeled only whole numbers, we can also find a place for numbers between the whole numbers. See Figure 2.

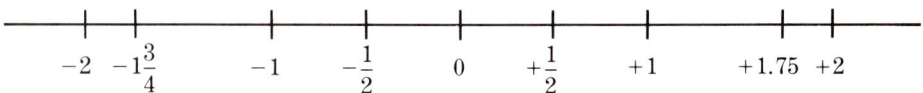

Figure 2

$+\dfrac{1}{2}$ is between 0 and $+1$

$-\dfrac{1}{2}$ is between -1 and 0

$+1.75$ is between $+1$ and $+2$

$-1\dfrac{3}{4}$ is between -2 and -1

In Figure 3 we show a number line with some numbers marked.

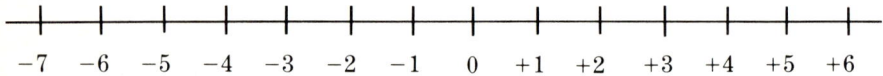

Figure 3

The numbers are in order of size and get larger as we go from left to right. So for any two numbers, the one on the right is larger.

$+5$ is to the right of -6, so $+5$ is larger than -6
-6 is to the right of -7, so -6 is larger than -7

These make sense when you remember that a temperature of $+5°$ is higher than a temperature of $-6°$. And a temperature of $-6°$ is higher than a temperature of $-7°$.

We call the numbers larger than zero **positive numbers**; we call the numbers smaller than zero **negative numbers**. Zero is not positive and it is not negative.

0 is to the left of any positive number, so 0 is smaller than any positive number.

0 is to the right of any negative number, so 0 is larger than any negative number.

We can often omit the sign of a positive number. Thus "$+5$" can be written as "5."

We can never omit the sign of a negative number. Thus "-5" must be written as "-5."

We call $+5$ and -5, the **opposites** of each other. So the opposite of $+5$ is -5, and the opposite of -5 is $+5$. The opposite of -1 is $+1$, and the opposite of 1 is -1. The opposite of 0 is 0.

Addition of Signed Numbers

If the temperature is +75 degrees, and it rises (increases) 4 degrees, then the new temperature is +79 degrees. See Figure 4.

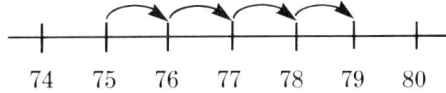

Figure 4

So   (+75) + (+4) = +79, or for short, 75 + 4 = 79.

If the temperature is −10 degrees, and it rises (increases) 4 degrees, the new temperature will be −6 degrees. See Figure 5.

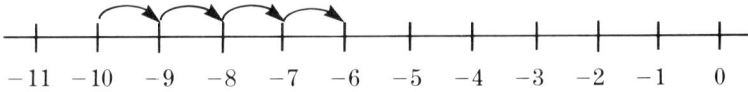

Figure 5

So   (−10) + (+4) = −6, or for short, −10 + 4 = −6. We add +4 to −10 by starting at −10 and moving 4 to the right. Thus, adding −10 and +4 gives you −6.

If the temperature is −3 degrees, and it rises (increases) 3 degrees, the new temperature will be 0 degrees. See Figure 6.

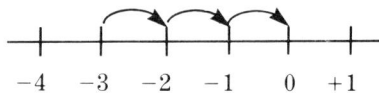

Figure 6

So   (−3) + (+3) = 0, or for short, −3 + 3 = 0. We add +3 to −3 by starting at −3 and moving 3 to the right. Thus adding −3 and +3 gives you 0.

If the temperature is −4 degrees, and it rises (increases) 6.5 degrees, the new temperature will be +2.5 degrees. See Figure 7.

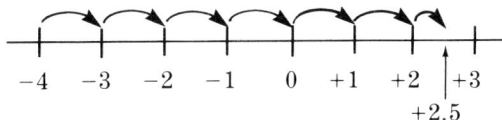

Figure 7

So $(-4) + (6.5) = +2.5$, or for short, $-4 + 6.5 = +2.5$. We add $+6.5$ to -4 by starting at -4 and moving 6.5 to the right. Thus, adding -4 and $+6.5$ gives you $+2.5$.

If the temperature is $+2$ degrees, and it falls (decreases) $5\frac{1}{2}$ degrees, the new temperature will be $-3\frac{1}{2}$ degrees. See Figure 8.

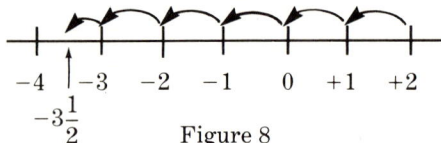

Figure 8

So $(+2) + (-5\frac{1}{2}) = -3\frac{1}{2}$, or for short, $+2 - 5\frac{1}{2} = -3\frac{1}{2}$. We add $-5\frac{1}{2}$ to $+2$ by starting at $+2$ and moving $5\frac{1}{2}$ units to the left. Thus, adding $+2$ and $-5\frac{1}{2}$ gives you $-3\frac{1}{2}$.

If the temperature is -3 degrees, and it falls (decreases) 4 degrees, the new temperature is -7 degrees. See Figure 9.

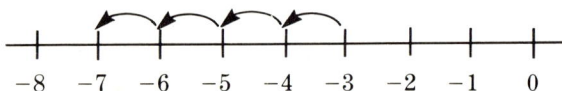

Figure 9

So $(-3) + (-4) = -7$, or for short, $-3 - 4 = -7$. We add -4 to -3 by starting at -3 and moving 4 units to the left. Thus, adding -4 to -3 gives you -7.

We have just done the following addition problems:

$$
\begin{aligned}
+75 &+ 4 &&= +79 \\
-10 &+ 4 &&= -6 \\
-3 &+ 3 &&= 0 \\
-4 &+ 6.5 &&= +2.5 \\
+2 &+ (-5\tfrac{1}{2}) &&= -3\tfrac{1}{2} \\
-3 &+ (-4) &&= -7
\end{aligned}
$$

Here are some more examples.

$$
\begin{aligned}
5.4 &+ 8.3 &&= +13.7 \\
5\tfrac{1}{2} &+ 8\tfrac{1}{2} &&= +14 \\
-7 &+ (-4) &&= -11 \\
-3.1 &+ (-2.1) &&= -5.2 \\
-3 &+ 0 &&= -3
\end{aligned}
$$

$$
\begin{array}{rcll}
5 & + & (-3) & = +2 \\
-5 & + & 3 & = -2 \\
8.7 & + & (-2) & = +6.7 \\
2 & + & (-8) & = -6 \\
-8 & + & 2 & = -6 \\
+5\tfrac{7}{8} & + & (-5\tfrac{7}{8}) & = \ \ 0 \\
+6 & + & 0 & = +6
\end{array}
$$

We can change the order when we add and still get the same answer.

$$
\begin{array}{rcrlrcl}
\text{So} \quad 5 & + & 8 & = 13 & \text{and} \quad 8 + 5 & = & 13 \\
-7.6 & + & 3 & = -4.6 & \text{and} \quad 3 + -7.6 & = & -4.6 \\
-4 & + & 4 & = 0 & \text{and} \quad 4 + (-4) & = & 0 \\
8 & + & (-8) & = 0 & \text{and} \quad -8 + 8 & = & 0 \\
-1 & + & 1 & = 0 & \text{and} \quad 1 + (-1) & = & 0
\end{array}
$$

Note: We always get zero when we add a number and its opposite.

To add more than two signed numbers, we add them two at a time in any convenient order.

EXAMPLE 1   Do the addition problem: $-3 + 5 + (-1)$

This can be done from left to right.

$$
\begin{array}{c}
-3 + 5 + (-1) \\
\underbrace{} \\
+2
\end{array}
$$

Now we add $+2$ and -1.

$$+2 + (-1) = +1$$

So $-3 + 5 + (-1) = +1.$

EXAMPLE 2   Do the addition problem: $-6 + 2 + 3 + (-8)$

We can add in any order and get

$$-6 + 2 + 3 + (-8) = -9$$

EXAMPLE 3   Do the addition problem: $2.1 + 3.2 + (-5) + (-3.4)$

We add the positive numbers and then we add the negative numbers.

$$2.1 + 3.2 + (-5) + (-3.4)$$

$$+5.3 \qquad\qquad -8.4$$

Now we add the two results.

$$5.3 + (-8.4) = -3.1$$

So $2.1 + 3.2 + (-5) + (-3.4) = -3.1.$

EXAMPLE 4   Do the addition problem: $10 - 5 - 2$

$$10 - 5 - 2$$

is shorthand for $10 + (-5) + (-2)$

and $10 + (-5) + (-2) = 3.$

EXAMPLE 5   Do the addition problem: $2.1 + 3.2 - 5 - 3.4$

$$2.1 + 3.2 - 5 - 3.4$$

is shorthand for $2.1 + 3.2 + (-5) + (-3.4)$

This is the same as example 3, so the answer is $-3.1.$

Subtraction of Signed Numbers

We know that

$$(10) - (3) = 7$$

But we also know that

$$(10) + (-3) = 7$$

Also $(14) - (9) = 5$ and $(14) + (-9) = 5$
   $(12) - (1) = 11$ and $(12) + (-1) = 11$
   $(24) - (20) = 4$ and $(24) + (-20) = 4$

The examples on the left are subtraction problems; the examples on the right are addition problems. In each example the subtraction problem on the left gives the same answer as the addition problem on the right. The only difference is that the second number in the left hand example is the opposite of the second number in the right hand example.

$$-9 \quad \text{is the opposite of } 9$$
$$-1 \quad \text{is the opposite of } 1$$
$$-20 \quad \text{is the opposite of } 20$$

So it is not necessary to learn new methods for subtraction. We get the same answer if instead of subtracting, we add the opposite of the second number.

EXAMPLE 6

Subtraction problem	becomes	addition problem	
$(-9) - (11)$		$(-9) + (-11)$	$= -20$
$(+4) - (12)$		$(+4) + (-12)$	$= -8$
$(+3.2) - (-4.7)$		$(+3.2) + (+4.7)$	$= +7.9$
$(0) - (5.25)$		$(0) + (-5.25)$	$= -5.25$
$(-3) - (-2)$		$(-3) + (+2)$	$= -1$
$(-3) - (0)$		$(-3) + (0)$	$= -3$

Note: In the last example, zero is its own opposite.

Subtraction problems are sometimes written with words in place of some of the symbols. For example,

$$\left.\begin{array}{l} \text{"subtract } -7 \text{ from 2"} \\[1em] \text{"from 2 subtract } -7\text{"} \end{array}\right\} \text{ both mean } (+2) - (-7)$$

To remember how to translate from words into symbols, notice that the number after the word "from" is written first.

EXAMPLE 7   Subtract -7 from 2.

Subtraction problem: $(+2) - (-7)$

Instead of subtracting, we add the opposite of the second number.

The opposite of -7 is $+7$.

Subtraction problem: $(+2) - (-7)$

becomes addition problem: $(+2) + (+7)$

But $(+2) + (+7) = 9$

So $(+2) - (-7) = 9$. Thus, when we subtract -7 from 2 we get 9.

EXAMPLE 8   From -8 subtract 3.

Subtraction problem: $(-8) - (+3)$

We change this to an addition problem as follows:

The opposite of $+3$ is -3.

Subtraction problem: $(-8) - (+3)$

becomes addition problem: $(-8) + (-3)$

But $(-8) + (-3) = -11$

So $(-8) - (+3) = -11$.

Multiplication of Signed Numbers

There are several ways of writing the same multiplication problem:

$$
\left.
\begin{array}{c}
(3)(-2) \\
3(-2) \\
(+3)(-2) \\
+3(-2) \\
-2(+3) \\
-2(3) \\
(-2)(+3) \\
(-2)(3) \\
(-2)3
\end{array}
\right\}
\quad \text{All mean the same thing}
$$

When we multiply two numbers:

    If the signs are the same the answer is positive
    If the signs are different the answer is negative.

In the multiplication problem $(+3)(-2)$, the two numbers have <u>different</u> <u>signs</u> so the answer is <u>negative</u>.

$$(+3)\,(-2) = -6$$

In the multiplication problem $(-4)(+2)$, the two numbers have <u>different</u> <u>signs</u> so the answer is negative.

$$(-4)(+2) = -8$$

In the multiplication problem $(-4)(-3)$, the two numbers have the <u>same</u> <u>signs</u> so the answer is <u>positive</u>.

$$(-4)(-3) = +12$$

In the multiplication problem $(+3)(+2)$, the two numbers have the <u>same</u> <u>signs</u> so the answer is <u>positive</u>.

$$(+3)(+2) = +6$$

In the multiplication problem $(+3)(0)$, one of the numbers is 0, so the answer is 0.

$$(+3)(0) = 0$$

When we multiply by zero, the answer is always zero.

EXAMPLE 9   Do the following multiplication problems.

$$(-2)(-8) = 16$$
$$(2)(8) = 16$$
$$(-8)(-2) = 16$$
$$(-2)(8) = -16$$
$$(2)(-8) = -16$$
$$(8)(-2) = -16$$
$$(-8)(2) = -16$$

To multiply more than two signed numbers, we multiply them two at a time in any convenient order.

EXAMPLE 10   Do the multiplication problem: $(+3)(+2)(-4)$

$$(+3)(+2)(-4)$$
$$(+6)(-4)$$
$$-24$$

So $(+3)(+2)(-4) = -24.$

EXAMPLE 11   Do the multiplication problem: $(-3)(-1)(-2)(+5)$

$$(-3)(-1)(-2)(+5)$$
$$(+3) \qquad (-10)$$
$$-30$$

So $(-3)(-1)(-2)(+5) = -30.$

EXAMPLE 12   Do the multiplication problem: $(-2.3)(-5.4)(-2)$

$$(-2.3)(-5.4)(-2)$$
$$(+12.42)(-2)$$
$$-24.84$$

So $(-2.3)(-5.4)(-2) = -24.84.$

EXAMPLE 13   Do the multiplication problem: $\left(-\dfrac{1}{3}\right)\left(+\dfrac{2}{5}\right)\left(-\dfrac{7}{3}\right)$

$$\left(-\frac{1}{3}\right)\left(+\frac{2}{5}\right)\left(-\frac{7}{3}\right)$$
$$\left(-\frac{2}{15}\right)\left(-\frac{7}{3}\right)$$
$$+\frac{14}{45}$$

So $\left(-\dfrac{1}{3}\right)\left(+\dfrac{2}{5}\right)\left(-\dfrac{7}{3}\right) = +\dfrac{14}{45}.$

Division of Signed Numbers

The rule for division of signed numbers is the same as the rule for multiplication of signed numbers.

> When we divide two numbers:
>
> If the <u>signs are the same</u>, the answer is <u>positive</u>.
> If the <u>signs are different</u>, the answer is <u>negative</u>.
> Dividing by zero has no meaning; that is, division by zero is undefined.

A division problem can be written as a fraction. For example,

$$(+12) \div (-2) \text{ is the same as } \frac{+12}{-2}$$

In the division problem $\dfrac{+12}{-2}$ the two numbers have <u>different signs</u>, so

the answer is <u>negative</u>. Thus, $(+12) \div (-2)$ or $\dfrac{+12}{-2} = -6$.

EXAMPLE 14 Divide -12 by -4.

The signs of the two numbers are the same, so the answer is positive.

$$\frac{-12}{-4} = +3$$

EXAMPLE 15 Divide 4.25 by -5.

The signs of the two numbers are different, so the answer is negative.

$$\frac{4.25}{-5} = -0.85$$

EXAMPLE 16 Do the problem: $-33 \div 2.2$

The signs of the two numbers are different, so the answer is negative. We use long division.

$$
\begin{array}{r}
1\,5. \\
2.2\,\overline{\smash{)}\,33.0} \\
22 \\
\hline
11\,0 \\
11\,0 \\
\hline
0
\end{array}
$$

So $-33 \div 2.2 = -15.$

Exercises

Find the larger number in each pair:
1. 6 and 3
2. −7 and −5
3. 0 and 2
4. 8 and −8
5. −2 and 1
6. $-\dfrac{3}{4}$ and $-\dfrac{1}{4}$

Answer True or False:
7. 7 is smaller than 5
8. −7 is smaller than −5
9. 0 is larger than 2
10. 0 is larger than −2
11. −5 is smaller than 5
12. −5 is smaller than −9
13. −5 is smaller than 9
14. 1.5 is smaller than 1

Add the following numbers:
15. $5 + 7$
16. $6 + (-2)$
17. $-4 + (-8)$
18. $4 + (-8)$
19. $-4 + 8$
20. $4 + 8$
21. $-8 + 0$
22. $-5 + 7.4$
23. $-5 + (-2.3)$
24. $-5 + 5.4$
25. $3 + \left(-8\dfrac{1}{2}\right)$
26. $-3 + \left(-8\dfrac{1}{2}\right)$

Add the following numbers:
27. $3 + 6 + 2$
28. $5 + 2 + (-3)$
29. $4 - 3 + 7$
30. $-4 + 5 - 2$
31. $9 + (-11) + 2$
32. $-4 + (-3) + (-1)$
33. $-4 - 3 - 1$

$-4 + (-3) + (-1)$
$-7 + 1 - 8$

34. $7 + 5 - 5 + 3$
35. $-2 - 7 + 4 - 1 + 2$
36. $-1 - 2 - 3 - 4$
37. $9 - 2 + 2 - 9$
38. $3.1 + 4.2 + (-5)$
39. $6.2 + (-2.1) + 3.3$
40. $4.2 + 2.1 + (-4) + (-4.5)$
41. $1.2 + 2.4 - 3.2 - 4.7$

Give the opposite of each of the following.
42. -19
43. $+19$
44. 0
45. -1.7
46. $\dfrac{2}{3}$

Do the following subtraction problems.
47. Subtract 5 from 8
48. Subtract 3 from 9
49. Subtract 6 from 2
50. Subtract 7 from 4
51. Subtract -5 from 8
52. Subtract 3 from -9
53. Subtract -6 from -2
54. Subtract -3 from -10
55. Subtract 0 from -2
56. Subtract 0 from 2
57. Subtract 2 from 0
58. Subtract -2 from 0
59. Subtract -4.2 from 0
60. Subtract 0 from 5.5
61. Subtract $(2 + 4)$ from 3
62. Subtract $(-2 - 3)$ from 8
63. Subtract $(7 - 8 + 2)$ from $(-3 + 5)$
64. From 10 subtract 2
65. From 7 subtract -2
66. From -7 subtract -2
67. From -2 subtract 7
68. From -2 subtract -7
69. From 2 subtract 7
70. From 2 subtract -7
71. From 4.6 subtract 2.2
72. From -5.3 subtract 3.1
73. From -6.7 subtract -1.6

Do the following multiplication problems.

74. $(+2)(+3)$
75. $(+3)(-4)$
76. $(-5)(+2)$
77. $(-6)(-5)$
78. $(-1)(+1)$
79. $(4)(-4)$
80. $(0)(-3)$
81. $(-7)(-2)$
82. $(-2)(5)$
83. $(3)(2)(1)$
84. $(-4)(1)(3)$
85. $(-2)(-5)(2)$
86. $(-4)(-2)(-1)$
87. $(5)(-4)(2)$
88. $(-2)(3)(0)$
89. $(1)(-7)$
90. $(2)(-3)(-5)$
91. $(-4)(-2.2)$
92. $(-3.1)(-2.5)(-2)$
93. $\left(-\dfrac{2}{3}\right)\left(+\dfrac{4}{5}\right)$
94. $\left(+\dfrac{1}{2}\right)\left(-\dfrac{3}{4}\right)\left(-\dfrac{5}{7}\right)$

Do the following division problems.

95. $\dfrac{-10}{+5}$

96. $\dfrac{10}{-5}$

97. $\dfrac{-10}{-5}$

98. $\dfrac{-8}{+2}$

99. $\dfrac{8}{-2}$

100. $\dfrac{-8}{-2}$

101. $\dfrac{-6}{4}$

102. Divide −15 by −5
103. Divide −1.5 by −5
104. Divide 8.4 by −4
105. Divide −42 by 1.2

Answers to Exercises

① 6

② -5

③ 2

④ 8

⑤ 1

⑥ $-\frac{1}{4}$

⑦ False

⑧ True

⑨ False

⑩ True

⑪ True

⑫ False

⑬ True

⑭ False

⑮ 5+7 is +12

⑯ 6+(-2) is +4

⑰ -4+(-8) is -12

⑱ 4 +(-8) is -4

⑲ -4+8 is +4

⑳ 4+8 is +12

㉑ -8+0 is -8

㉒ -5+7.4 is +2.4

㉓ -5+(-2.3) is -7.3

㉔ -5+5.4 is +0.4

㉕ 3+(-8½) is -5½

㉖ -3+(-8½) is -11½

㉗ $\underbrace{3+6}_{}+2$ is
$\underbrace{9+2}_{11}$ is

㉘ $\underbrace{5+2}_{}+(-3)$ is
$\underbrace{7+(-3)}_{+4}$ is

㉙ 4-3+7 becomes
$\underbrace{(4)+(-3)}_{(+1)}+(7)$ is
$\underbrace{(+1)+(7)}_{+8}$, or

㉚ -4+5-2 becomes
$\underbrace{(-4)+(5)}_{(+1)}+(-2)$ is
$\underbrace{(+1)+(-2)}_{(-1)}$, or

㉛ $\underbrace{9+(-11)}_{}+2$ is
$\underbrace{-2+2}_{0}$ is

㉜ $\underbrace{-4+(-3)}_{(-7)}+(-1)$ is
$\underbrace{(-7)+(-1)}_{-8}$

㉝ $-4-3-1$ becomes
$(-4)+(-3)+(-1)$ or
$\underbrace{-7 \; + \; (-1)}$, or
-8

㉞ $7+5-5+3$ becomes
$\underbrace{(7)+(5)} + \underbrace{(-5)+(3)}$, is
$\underbrace{+12 \; + \; (-2)}$ or
$+10$

㉟ $-2-7+4-1+2$ becomes
$\underbrace{(-2)+(-7)}+\underbrace{(+4)+(-1)}+(+2)$
$\underbrace{(-9) \; + \; (+3)}, +(+2)$
$\underbrace{(-6) \quad +2}$, or
(-4)

㊱ $-1-2-3-4$ becomes
$\underbrace{(-1)+(-2)} + \underbrace{(-3)+(-4)}$ or
$\underbrace{(-3) \; + \; (-7)}$ is
-10

㊲ $9-2+2-9$ becomes
$\underbrace{(+9)+(-2)} + \underbrace{(+2)+(-9)}$ is
$\underbrace{(+7) \quad + \quad (-7)}$, or
0

㊳ $3.1+4.2+(-5)$ becomes
$\underbrace{(3.1)+(4.2)}+(-5)$ is
$\underbrace{(+7.3) \quad + \quad (-5)}$ or
$+2.3$

㊴ $6.2+(-2.1)+3.3$ becomes
$\underbrace{(6.2)+(-2.1)}+(3.3)$ is
$\underbrace{(+4.1) \; + \; (3.3)}$ or
$+7.4$

40. $4.2 + 2.1 + (-4) + (-4.5)$

becomes $\underline{(4.2) + (2.1)} + \underline{(-4) + (-4.5)}$, is

$$(+6.3) + (-8.5)$$
$$or \quad -2.2$$

41. $1.2 + 2.4 - 3.2 - 4.7$ becomes

$\underline{(1.2) + (2.4)} + \underline{(-3.2) + (-4.7)}$, is

$$(+3.6) + (-7.9) \quad or$$
$$-4.3$$

42. $+19$

43. -19

44. 0

45. $+1.7$

46. $-\frac{2}{3}$

47. $(8) - (5)$ becomes
$(8) + (-5)$ is $+3$

48. $(9) - (3)$ becomes
$(9) + (-3)$ is $+6$

49. $(2) - (6)$ becomes
$(2) + (-6)$ or (-4)

50. $(4) - (7)$ becomes
$(4) + (-7)$ or -3

51. $8 - (-5)$ becomes
$(8) + (+5)$ or $+13$

52. $(-9) - (3)$ becomes
$(-9) + (-3)$ or -12

53. $(-2)-(-6)$ becomes
$(-2)+(+6)$ or $+4$

54. $-10-(-3)$ becomes
$(-10)+(+3)$ or (-7)

55. $-2-0$ becomes
$-2+0$ or -2

56. $2-0$ becomes
$2+0$ or 2

57. $0-2$ becomes
$0+(-2)$ or -2

58. $0-(-2)$ becomes
$0+(+2)$ or $+2$

59. $0-(-4.2)$ becomes
$0+(+4.2)$ or
$+4.2$

60. $5.5-0$ becomes
$5.5+0$ or 5.5

61. $3-(2+4)$ becomes
$3-(+6)$ or
$3+(-6)$ is -3

62. $8-(-2-3)$ becomes
$8-((-2)+(-3))$ or
$8-(-5)$ which becomes
$8+5$ is 13

63. $(-3+5)-(7-8+2)$
$(+2)-(7+(-8)+2)$
$(+2)-(1)$ becomes
$(+2)+(-1)$ is $+1$

64. $10-2$ becomes
$10+(-2)$ or $+8$

65. $7-(-2)$ becomes
$7+(+2)$ or $+9$

66. $-7-(-2)$ becomes
$-7+(+2)$ or -5

67. $-2-(7)$ becomes
$-2+(-7)$ or -9

68. $-2-(-7)$ becomes
$(-2)+(+7)$ or $+5$

69. $2-(7)$ becomes
$2+(-7)$ or -5

70. $2-(-7)$ becomes
$2+(+7)$ or $+9$

71. $4.6-(2.2)$ becomes
$4.6+(-2.2)$ or 2.4

72. $-5.3-(3.1)$ becomes
$-5.3+(-3.1)$ or -8.4

73. $-6.7-(-1.6)$ becomes
$-6.7+(+1.6)$ or -5.1

74. $(+2)(+3)$ is $+6$

75. $(+3)(-4)$ is -12

76. $(-5)(+2)$ is -10

77. $(-6)(-5)$ is $+30$

78. $(-1)(+1)$ is -1

79. $(4)(-4)$ is -16

80. $(0)(-3)$ is 0

81. $(-7)(-2)$ is $+14$

82. $(-2)(5)$ is -10

83. $\underline{(3)(2)}\,(1)$ is
$(6)\ \ (1)$ or 6

84. $\underline{(-4)\,(1)}\,(3)$
$(-4)\ (3)$ or -12

85.) $(-2)(-5)(2)$ is
$(+10)$ (2) or $+20$

86.) $(-4)(-2)(-1)$ is
$(+8)$ (-1) or -8

87.) $(5)(-4)(2)$ is
$(-20)(2)$ or -40

88.) $(-2)(3)(0)$ is
(-6) (0) or 0

89.) $(1)(-7)$ is -7

90.) $(2)(-3)(-5)$ is
(-6) (-5) or
$+30$

91.) $(-4)(-2.2)$ is $+8.8$

92.) $(-3.1)(-2.5)(-2)$ is
$(+7.75)$ (-2) or
-15.5

93.) $\left(-\frac{2}{3}\right)\left(+\frac{4}{5}\right)$ is $-\frac{8}{15}$

94.) $\left(+\frac{1}{2}\right)\left(-\frac{3}{4}\right)\left(-\frac{5}{7}\right)$ is
$\left(-\frac{3}{8}\right)\left(-\frac{5}{7}\right)$ or $+\frac{15}{56}$

95.) $\frac{-10}{5}$ is $-\frac{2\cdot 5}{5}$ or -2

96.) $\frac{10}{-5}$ is $-\frac{2\cdot 5}{5}$ or -2

97.) $\frac{-10}{-5}$ is $+\frac{2\cdot 5}{5}$ or $+2$

98.) $\frac{-8}{+2}$ is $-\frac{4\cdot 2}{2}$ or -4

99.) $\frac{8}{-2}$ is $-\frac{4\cdot 2}{2}$ or -4

100.) $\frac{-8}{-2}$ is $+\frac{4\cdot 2}{2}$ or $+4$

101.) $\frac{-6}{4}$ is $-\frac{3\cdot 2}{2\cdot 2}$ or $-\frac{3}{2}$

102.) $\frac{-15}{-5}$ is $+\frac{3\cdot 5}{5}$ or $+3$

(103.) $\dfrac{-1.5}{-5}$ is $+0.3$

(104.) $\dfrac{8.4}{-4}$ is -2.1

(105.)

$$
\begin{array}{r}
35. \\
1.2\overline{)42.0.} \\
\underline{36} \\
6\ 0 \\
\underline{6\ 0} \\
0
\end{array}
$$

Additional Exercises

Find the larger number in each pair.
1. 10 and 7
2. −4 and −6
3. 3 and 0
4. 6 and −6
5. −5 and 2
6. $-\dfrac{1}{3}$ and $-\dfrac{2}{3}$

Answer <u>true</u> or <u>false</u>.
7. 4 is larger than 2
8. −4 is larger than −2.
9. 0 is smaller than 6
10. 0 is smaller than −6
11. −3 is smaller than 3
12. −3 is smaller than −8
13. −3 is smaller than 8
14. 1.7 is smaller than 1

Add the following numbers.
15. 6 + 1
16. 9 + (−2)
17. −5 + (−6)
18. 5 + (−6)
19. −3 + 7
20. 3 + 7
21. −7 + 0
22. −9 + 0
23. −4 + 8.2
24. −4 + (−3.7)
25. $5 + \left(-9\dfrac{1}{4}\right)$
26. $-5 + \left(-9\dfrac{1}{4}\right)$

Add the following numbers.
27. 5 + 1 + 6
28. 3 + 7 + (−2)
29. 6 − 2 + 3
30. −6 + 7 − 2
31. 8 + (−12) + 1
32. −5 + (−2) + (−3)
33. −5 − 2 − 3

34. $4 - 4 + 3 - 1$
35. $-3 - 9 + 5 - 2 + 3$
36. $-2 - 3 - 4 - 5$
37. $7 - 3 - 7 + 3$
38. $4.2 + 5.3 + (-6)$
39. $7.4 + (-3.1) + 2.3$
40. $5.2 + 2.6 + (-3) + (-3.3)$
41. $1.4 + 3.2 - 2.6 - 3.2$

Give the opposite of each of the following.
42. $+16$
43. -16
44. 0
45. 5.4
46. $-\dfrac{4}{7}$

Do the following subtraction problems.
47. Subtract 7 from 9
48. Subtract 4 from 10
49. Subtract 8 from 3
50. Subtract 6 from 1
51. Subtract -6 from 9
52. Subtract 2 from -10
53. Subtract -7 from -1
54. Subtract -5 from -9
55. Subtract 0 from -4
56. Subtract 0 from 4
57. Subtract 4 from 0
58. Subtract -4 from 0
59. Subtract -5.2 from 0
60. Subtract 0 from 4.7
61. Subtract $(3 + 5)$ from 2
62. Subtract $(-3 - 4)$ from 9
63. Subtract $(6 - 9 + 4)$ from $(-2 + 7)$
64. From 9 subtract 5
65. From 8 subtract 11
66. From -2 subtract 10
67. From -1 subtract 9
68. From -3 subtract 8
69. From 0 subtract 0
70. From 4 subtract 4
71. From 4.7 subtract 3.2
72. From -6.1 subtract 2.7
73. From -7.4 subtract -2.3

Do the following multiplication problems.

74. $(+4)(+5)$
75. $(-5)(+2)$
76. $(+6)(-3)$
77. $(-2)(-7)$
78. $(-3)(+3)$
79. $(5)(-5)$
80. $(-6)(0)$
81. $(-8)(-1)$
82. $(-3)(4)$
83. $(-5)(-3.3)$
84. $(4)(3)(1)$
85. $(-2)(2)(5)$
86. $(-3)(-6)(2)$
87. $(-1)(-2)(-3)$
88. $(7)(-2)(2)$
89. $(0)(4)(-6)$
90. $(2)(-9)(-1)$
91. $(-5)(2)(4)$
92. $(-4.2)(-3.1)(-10)$
93. $\left(-\dfrac{3}{4}\right)\left(\dfrac{5}{7}\right)$
94. $\left(+\dfrac{3}{8}\right)\left(-\dfrac{8}{7}\right)\left(+\dfrac{1}{2}\right)$

Do the following division problems.

95. $\dfrac{-12}{+6}$

96. $\dfrac{12}{-6}$

97. $\dfrac{-12}{-6}$

98. $\dfrac{+21}{-7}$

99. $\dfrac{-21}{7}$

100. $\dfrac{-21}{-7}$

101. $\dfrac{-9}{6}$

102.   Divide -20 by -4
103.   Divide -20 by -0.4
104.   Divide 9.3 by -3
105.   Divide -36 by 1.8

Lesson 2

Algebraic Notation
and Addition of Polynomials

Algebraic Notation

In this lesson and in the rest of the work you do in mathematics, science, engineering or technology, you will often use letters in place of numbers.

Suppose records cost $3 each. Then,

> 2 records cost (3)(2) dollars
> 5 records cost (3)(5) dollars
> 0 records cost (3)(0) dollars
> x records cost (3)(x) dollars

In the last line, x stands for the number of records and (3)(x) dollars is the cost of x records. Shorthand for (3)(x) is $3x$.

There is a shipping charge for each order of records.

> to Boston $2
> to Chicago $3
> to Los Angeles $4

The shipping charge is added to the cost of the records to get the total bill.

If you buy x records and ship them, your total bill will be

$(3x + 2)$ dollars to ship to Boston
$(3x + 3)$ dollars to ship to Chicago
$(3x + 4)$ dollars to ship to Los Angeles

We can let y stand for the number of dollars of the shipping charge.

$(3x + y)$ dollars is the total bill

Let's compute the total bill for a person who buys 5 records and ships them to Boston. That is, when x is 5 and y is 2.

$3x + y$	dollars
$(3(5) + 2)$	dollars
$(15 + 2)$	dollars
17	dollars

In this problem $3x$ means 3 times x. In algebra we don't use $\times$ to mean multiplication. It looks too much like the letter x. Here are other ways of showing multiplication.

$3x$
$3(x)$
$(3)(x)$ } all mean multiply 3 by x
$3 \cdot x$

xy means x times y
x means $(1)(x)$
$-x$ means $(-1)(x)$

$3xy$
$(3)(x)(y)$
$3(xy)$
$(3x)y$ } all mean 3 times x times y
$(3)(xy)$
$(3x)(y)$

The expression $3 + 2x$ can be confusing because you don't know whether the 2 goes with the 3 or with the x. That is, should you first multiply the 2 by the x or first add the 2 to the 3? To avoid this confusion, mathematicians have agreed that in this kind of problem we do the multiplication first. So $3 + 2x$ means that 2 is multiplied by x and then 3 is added.

If x stands for 4,

    then $3 + 2x$

    means $3 + 2(4)$

    or $3 + 8$

    which is 11.

If x stands for -5,

    then $3 + 2x$

    means $3 + 2(-5)$

    or $3 + (-10)$

    which is -7.

EXAMPLE 1   If x is 5 and y is -4, what does $-3x + 2y + 7$ stand for?

    If x is 5 and y is -4,

    then $-3x + 2y + 7$

    stands for $(-3)(5) + (2)(-4) + 7$

Note that we do the multiplication first.

                $-15 - 8 + 7$

    means $-23 + 7$

    or -16.

EXAMPLE 2   If x is -2 and y is 5, what does $-3x + 2y + 7$ stand for?

    If x is -2 and y is 5

    then $-3x + 2y + 7$

    stands for $-3(-2) + 2(5) + 7$

    or $+6 + 10 + 7$

    which is $+23$.

EXAMPLE 3   If x is 1.5 and y is -0.7, what does $-3x + 2y + 7$ stand for?

If x is 1.5 and y is -0.7

then $-3x + 2y + 7$

stands for $-3(1.5) + 2(-0.7) + 7$

or $-4.5 -$ 1.4 $+ 7$

which is $+1.1$.

EXAMPLE 4   Write an expression that shows the total bill for buying several rolls of film that cost \$4.25 each and several records that cost \$2.98 each.

$4.25\ x$

Let's use the letter f to stand for the number of rolls of film we are buying. Then the cost of the film is

$4.25f$

Now we need a letter to stand for the number of records we are buying. We can use any letter except f (because f already stands for the number of rolls of film). Let's use the letter r. Then the cost of the records is

$2.98r$

So our total bill in dollars for f rolls of film and r records is

$4.25f + 2.98r$

EXAMPLE 5   Use the expression we got in example 4 to find the total bill for buying 3 rolls of film and 4 records.

The expression for the cost of f rolls of film and r records is

$4.25f + 2.98r$

When f is 3 and r is 4, this expression becomes

$4.25(3) + 2.98(4)$
$12.75 + 11.92$
24.67

So our total bill will be \$24.67.

Polynomial Expressions

The expression $-3x + 2y + 7$ consists of <u>three</u> terms: $-3x$, $+2y$ and $+7$.

The expression $2x + y - \widecirc{xy} - 2$ consists of <u>four</u> terms: $2x$, $+y$, $-xy$ and -2.

The expression $5x$ consists of one term.

We call an expression with one term a **monomial** ("mono" means one).
We call an expression with two terms a **binomial**) ("bi" means two).
We call an expression with three terms a **trinomial** ("tri" means three).
<u>We call any expression with more than one term</u> a **polynomial** ("poly" means many).

Adding in Algebra

Remember $3(2)$ means $2 + 2 + 2$

So $3(x)$ means $x + x + x$

and $3x + 2x$ means $(x + x + x) + (x + x)$
                                which is $x + x + x + x + x$
                                or $5x$.

Note that the problem $3x + 2x = 5x$ is very much like the problem $3 + 2 = 5$.

Look at the pattern in each of the following examples.

$$3 + 2 = 5 \quad \text{and} \quad 3x + 2x = 5x$$
$$+2 + 7 = +9 \quad \text{and} \quad +2x + 7x = +9x$$
$$-3 + 4 = +1 \quad \text{and} \quad -3y + 4y = +1y \text{ or } y$$
$$3 - 4 = -1 \quad \text{and} \quad 3y - 4y = -1y \text{ or } -y$$
$$-7 - 3 = -10 \quad \text{and} \quad -7a - 3a = -10a$$
$$-4 + 4 = 0 \quad \text{and} \quad -4b + 4b = 0$$

same letters

like terms

Terms that have <u>exactly</u> the same letters are called **like terms**. In each of the examples above we simplified the expression by adding like terms. <u>Notice that we add like terms in the same way we add signed numbers.</u>

EXAMPLE 6   Simplify each of the following expressions by adding like terms.

$3x + x = 4x$ Remember x means $1x$
$3y - y = 2y$ Remember $-y$ means $-1y$
$7b + 2b - 5b = 4b$
$-3a + 5a - 2a = 0a$ which is 0
$40s - 70s - 30s = -60s.$

EXAMPLE 7   Simplify the expression $4x - 2x + 3y$ by adding like terms.

$4x$ and $-2x$ are like terms so we can add them.

$$4x - 2x + 3y$$

becomes $2x + 3y.$

Watch it! $2x + 3y$ means $(x + x) + (y + y + y)$, which is two x's and three y's, and there is no way of combining $2x$ and $3y$ into a single term. So when we add $2x$ and $3y$, we get $2x + 3y$. In an addition problem we can only combine like terms.

EXAMPLE 8   Simplify each of the following expressions by adding like terms.

$6x + 3x = 9x$
$2y + 3y = 5y$
$4x + x + 9y + 7y = 5x + 16y$
$10x + 2y + 5x = 15x + 2y$
$5xy - 2xy = 3xy$
$8xy + 2x = 8xy + 2x$

There is nothing we can do with this one because $8xy$ and $2x$ are not like terms. (They don't have exactly the same letters.)

$5a - 2b - 3a + 7b = 2a + 5b$
$4ab + 2a - 3b + 3ab - a + 4b = 7ab + a + b$
$2x - 3y + 7 - 5x + 10y + 2 = -3x + 7y + 9$
$2x + 3y - 4x + 7y = -2x + 10y$

EXAMPLE 9   Add the polynomial $2x + 3y$ to the polynomial $-4x + 7y$.

$$(2x + 3y) + (-4x + 7y)$$

is the same as $2x + 3y - 4x + 7y$

We add like terms and get

$$-2x + 10y$$

So $(2x + 3y) + (-4x + 7y) = -2x + 10y.$

EXAMPLE 10   Add the three polynomials $-x + 2y + 3$ and $2x + 7y$ and $-x - y - 2.$

$$(-x + 2y + 3) + (2x + 7y) + (-x - y - 2)$$

is the same as $-x + 2y + 3 + 2x + 7y - x - y - 2$

Grouping like terms we get

$$(-x + 2x - x) + (2y + 7y - y) + (3 - 2)$$

Now we add like terms and get

$$0x + 8y + 1$$

which is the same as $0 + 8y + 1$ Remember $0x$ is 0

So $(-x + 2y + 3) + (2x + 7y) + (-x - y - 2) = 8y + 1.$

We can add the polynomials in example 10 in another way. We line up like terms under each other and then add.

$$
\begin{array}{l}
-1x + 2y + 3 \\
\ \ 2x + 7y \\
\underline{-1x - 1y - 2} \\
\end{array}
$$

(*Notice:* We write $-x$ as $-1x$ and $-y$ as $-1y$.)

which is $0x + 8y + 1$

or $8y + 1.$

EXAMPLE 11   Add the two polynomials $2.5a + 4.7b - 6.1$ and $1.6a + 8.3.$

We line up like terms and then add.

$$
\begin{array}{l}
2.5a + 4.7b - 6.1 \\
\underline{1.6a \ \ \ \ \ \ \ \ \ + 8.3} \\
4.1a + 4.7b + 2.2
\end{array}
$$

Exercises

Do the following word problems.

1. If an object costs x dollars, then how much will 2 of them cost? $2x$

2. If an object costs y cents and the price is then raised 5 cents, what will the new cost be? $y + 5$

✓3. If an object sells for d dollars, how much will m of them cost?
$d(m) \implies m(d)$ dollars

When x is 5, what does each of the following stand for?

4. x 5
5. $2x$ 10
6. $6x$
7. $3x$
8. $-x$
9. $-5x$
10. $-6x$
11. $x + 1$
12. $x - 1$
13. $x + 3$
14. $2x + 3$
15. $3x + 2$ 17
16. $3 - 2x$

When x is -3, what does each of the following stand for?

17. x -3
18. $2x$ -6
19. $6x$ -18
20. $3x$
21. $-x$ 3
22. $-5x$
23. $-6x$
24. $x + 1$
25. $x - 1$
26. $x + 3$
27. $2x + 3$
28. $3x + 2$ 4
29. $3 - 2x$

When x is 2 and y is -4, what does each of the following stand for?

30. xy
31. $x + y$
32. $2x + y$
33. $-3xy$

34. $2x + 3y + xy$
35. $x + y + 5$

When x is 2.5 and y is -0.3, what does each of the following stand for?
36. $-2x + y + 4$
37. $3x - 2y - 1$
38. $-4x + 3y$
39. $-3x - 2y$

Do the following word problems.
40. Write an expression that shows the total bill for buying several rolls of film that cost $3.50 each and several records that cost $2.75 each.

41. Use the expression in exercise 40 to find the total bill for buying 2 rolls of film and 3 records.

42. Write an expression that shows the total bill for buying several blouses that cost $4.98 each and several skirts that cost $6.00 each.

43. Use the expression in exercise 42 to find the total bill for buying 4 blouses and 2 skirts.

Add the following monomials.
44. x and x
45. x and $2x$
46. x and y
47. $3x$ and $7x$
48. $5y$ and $-2y$
49. $2x$ and $3y$
50. $3c$ and $-c$
51. d and d
52. $2x$ and $-5x$
53. $-7y$ and $-2y$ and $4y$

Combine like terms to simplify each of the following polynomials.
54. $3x + 2x$
55. $7x + 2x + x$
56. $2x + 3x - 2x$
57. $6x - 3x - 7x$
58. $3x + 2x + y$
59. $2x + 3x + 5y + 7y$
60. $3a - 7a + 8b - 2b$
61. $7r - 2s + 3r + 5s$
62. $8x - 3y - 4x + 2x + 5$

63. $3x - 2y + 7 - 3x + 3y - 7$
64. $3x + 4y - 2xy - 2x + 6xy$

Add the following polynomials.
65. $x + 1$ and $2x + 3$
66. $3x - 5$ and $4x + 6$
67. $2x + y$ and $3x + 2y$
68. $2x + 5$ and $-3x - 15$
69. $2x + 3$ and $5x + 4$
70. $8x + 3y$ and $-2x + 4y$
71. $-a + b$ and $2b - a$
72. $6j - 4k$ and $2j + 5k$ and $-3j + 5k$
73. $a + b + d$ and $a - b + c$
74. $2x + 7xy - 3y$ and $-4x - 3y - 5xy$
75. $-2a + 4b$ and $3a - b$ and $5a + 3b$
76. $4.5x + 3.2y$ and $1.1x - 2.1y$
77. $3.4x + 2.7y + 5$ and $-1.2x - 7$
78. $5.5a - 4.2b$ and $-4.4a + 7.8b$
79. $2.7a + 3.8b$ and $4.6a - 2.1b - 4$

Add the following polynomials.
80. $2x + 3y - 7$
    $\underline{4x - 2y + 12}$

81. $3j - 4k + 6$
    $\underline{-4j + 5k - 1}$

82. $3xy - 4x + y - 1$
    $\underline{2xy + 7x + 5}$

83. $-2m - 4n + 6$
    $\underline{3m - 2n - 5}$

84. $2x + 3y$
    $-4x - 3y$
    $\underline{-4x + 2y}$

85. $3x - 4y - 7$
    $7x + 3$
    $-2x - y + 4$
    $\underline{6x + y}$

86. $3.2x + 4.3y + 2$
    $\underline{2.7x - 1.1y - 5}$

87. $4.4x - 2.7y$
    $\quad\quad\quad 3.8y - 8$
    $\underline{2.9x \quad\quad\quad + 3}$

Answers to Exercises

① $2x$ dollars

② $y + 5$ cents

③ m d dollars

④ $x = 5$

⑤ $2x = 2(5) = 10$

⑥ $6x = 6(5) = 30$

⑦ $3x = 3(5) = 15$

⑧ $-x = -1(5) = -5$

⑨ $-5x = -5(5) = -25$

⑩ $-6x = -6(5) = -30$

⑪ $x + 1$ is
$5 + 1$ or 6

⑫ $x - 1$
$5 - 1$
4

⑬ $x + 3$
$5 + 3$
8

⑭ $2x + 3$
$2(5) + 3$
$10 + 3$
13

⑮ $3x + 2$
$3(5) + 2$
$15 + 2$
17

⑯ $3 - 2x$
$3 - 2(5)$
$3 - 10$
-7

⑰ $x = -3$

⑱ $2x$
$2(-3)$
-6

⑲ $6x$
$6(-3)$
-18

⑳ $3x$
$3(-3)$
-9

㉑ $-x$
$-1(-3)$
$+3$

㉒ $-5x$
$-5(-3)$
$+15$

㉓ $-6x$
$-6(-3)$
$+18$

㉔ $x + 1$
$-3 + 1$
-2

㉕ $x - 1$
$-3 - 1$
-4

㉖ $x + 3$
$-3 + 3$
0

㉗ $2x + 3$
$2(-3) + 3$
$-6 + 3$
-3

㉘ $3x + 2$
$3(-3) + 2$
$-9 + 2$
-7

㉙ $3 - 2x$
$3 - 2(-3)$
$3 + 6$
9

㉚ xy
$2(-4)$
-8

㉛ $x + y$
$2 + (-4)$
$(2) + (-4)$
-2

㉜ $2x + y$
$2(2) + (-4)$
$4 + (-4)$
0

㉝ $-3xy$
$-3(2)(-4)$
$(-6)(-4)$
$+24$

㉞ $2x + 3y + xy$
$2(2) + 3(-4) + (2)(-4)$
$4 + (-12) + (-8)$
$(-8) + (-8)$
-16

㉟ $x + y + 5$
$(2) + (-4) + 5$
$-2 + 5$
$+3$

㊱ $-2x + y + 4$
$-2(2.5) + (-0.3) + 4$
$-5 + (-0.3) + 4$
$-5.3 + 4$
-1.3

㊲ $3x - 2y - 1$
$3(2.5) - 2(-0.3) - 1$
$7.5 + 0.6 - 1$
$8.1 - 1$
7.1

38.) $-4x + 3y$
$-4(2.5) + 3(-0.3)$
$-10 \quad -0.9$
$\qquad -10.9$

39.) $-3x \quad -2y$
$-3(2.5) - 2(-0.3)$
$-7.5 + 0.6$
$\qquad -6.9$

40.) $3.50f + 2.75r$

41.) $3.50f + 2.75r$
$3.50(2) + 2.75(3)$
$7.00 + 8.25$
$\qquad 15.25$
total bill is # 15.25

42.) $4.98b + 6.00s$

43.) $4.98b + 6.00s$
$4.98(4) + 6.00(2)$
$19.92 + 12.00$
$\qquad 31.92$
total bill is # 31.92

44.) $x + x = 2x$

45.) $x + 2x = 3x$

46.) $x + y$

47.) $3x + 7x = 10x$

48.) $5y + (-2y) = +3y$

49.) $2x + 3y$

50.) $3c + (-c) = 2c$

51.) $d + d = 2d$

52.) $2x + (-5x) = -3x$

53.) $\underbrace{(-7y) + (-2y)}_{-9y} + \underbrace{(4y)}_{+4y}$
$\qquad -5y$

54.) $3x + 2x = 5x$

55.) $\underbrace{7x + 2x}_{9x} + \underbrace{x}_{+1x}$
$\qquad 10x$

56.) $2x + 3x - 2x$

$\underbrace{5x \quad - 2x}$

$3x$

57.) $6x - 3x - 7x$

$\underbrace{3x \quad - 7x}$

$-4x$

58.) $\underbrace{3x + 2x} + y$

$5x + y$

59.) $\underbrace{2x + 3x} \underbrace{+ 5y + 7y}$

$5x + 12y$

60.) $\underbrace{3a - 7a} \underbrace{+ 8b - 2b}$

$-4a + 6b$

61.) $7r - 2s + 3r + 5s$

$\underbrace{7r + 3r} \underbrace{- 2s + 5s}$

$10r + 3s$

62.) $8x - 3y - 4x + 2x + 5$

$\underbrace{8x - 4x} + 2x - 3y + 5$

$\underbrace{4x + 2x} - 3y + 5$

$6x - 3y + 5$

63.) $3x - 2y + 7 - 3x + 3y - 7$

$\underbrace{3x - 3x} \underbrace{- 2y + 3y} \underbrace{+ 7 - 7}$

$0 \quad + \quad y \quad + 0$

y

64.) $3x + 4y - 2xy - 2x + 6xy$

$\underbrace{3x - 2x} \underbrace{- 2xy + 6xy} + 4y$

$x \quad + \quad 4xy \quad + 4y$

$x + 4y + 4xy$

65.) $(x+1) + (2x+3)$

$x + 1 + 2x + 3$

$\underbrace{x + 2x} \underbrace{+ 1 + 3}$

$3x + 4$

66.) $(3x - 5) + (4x + 6)$

$3x - 5 + 4x + 6$

$\underbrace{3x + 4x} \underbrace{- 5 + 6}$

$7x + 1$

67.) $(2x + y) + (3x + 2y)$

$2x + y + 3x + 2y$

$\underbrace{2x + 3x} \underbrace{+ y + 2y}$

$5x + 3y$

68. $(2x+5)+(-3x-15)$
$2x+5-3x-15$
$\underbrace{2x-3x}+\underbrace{5-15}$
$-x-10$

69. $(2x+3)+(5x+4)$
$2x+3+5x+4$
$\underbrace{2x+5x}+\underbrace{+3+4}$
$7x+7$

70. $(8x+3y)+(-2x+4y)$
$8x+3y-2x+4y$
$\underbrace{8x-2x}+\underbrace{3y+4y}$
$6x+7y$

71. $(-a+b)+(2b-a)$
$-a+b+2b-a$
$\underbrace{-a-a}+\underbrace{b+2b}$
$-2a+3b$

72. $(6j-4k)+(2j+5k)+(-3j+5k)$
$6j-4k+2j+5k-3j+5k$
$\underbrace{6j+2j-3j}\ \underbrace{-4k+5k+5k}$
$\underbrace{8j-3j}+\underbrace{1k+5k}$
$5j+6k$

73. $(a+b+d)+(a-b+c)$
$a+b+d+a-b+c$
$\underbrace{a+a}+\underbrace{b-b}+c+d$
$2a+0+c+d$
$2a+c+d$

74. $(2x+7xy-3y)+(-4x-3y-5xy)$
$2x+7xy-3y-4x-3y-5xy$
$\underbrace{2x-4x}\ \underbrace{3y-3y}+\underbrace{7xy-5xy}$
$-2x-6y+2xy$

75. $(-2a+4b)+(3a-b)+(5a+3b)$
$-2a+4b+3a-b+5a+3b$
$\underbrace{-2a+3a}+5a\ \underbrace{+4b-b}+3b$
$\underbrace{1a+5a}+\underbrace{3b+3b}$
$6a+6b$

76. $(4.5x+3.2y)+(1.1x-2.1y)$
$4.5x+3.2y+1.1x-2.1y$
$\underbrace{4.5x+1.1x}+\underbrace{3.2y-2.1y}$
$5.6x+1.1y$

77. $(3.4x+2.7y+5)+(-1.2x-7)$
$3.4x+2.7y+5-1.2x-7$
$\underbrace{3.4x-1.2x}+\underbrace{2.7y}+\underbrace{5-7}$
$2.2x+2.7y-2$

78.) $(5.5a - 4.2b) + (-4.4a + 7.8b)$
$5.5a - 4.2b - 4.4a + 7.8b$
$\underbrace{5.5a - 4.4a} , \underbrace{-4.2b + 7.8b}$

$\qquad 1.1a \quad + \quad 3.6b$

84.) $\quad 2x + 3y$
$\quad -4x - 3y$
$\quad \underline{-4x + 2y}$
$\quad -6x + 2y$

79.) $(2.7a + 3.8b) + (4.6a - 2.1b - 4)$
$2.7a + 3.8b + 4.6a - 2.1b - 4$
$\underbrace{2.7a + 4.6a} + \underbrace{3.8b - 2.1b} , -4$

$\qquad 7.3a \quad + \quad 1.7b \quad -4$

85.) $3x - 4y \quad -7$
$\quad 7x \qquad\quad +3$
$\quad -2x \; -y \; +4$
$\quad \underline{6x \; +y}$
$\quad 14x - 4y + 0$

$\qquad$ or $14x - 4y$

80.) $\quad 2x + 3y \; -7$
$\quad \underline{4x \; -2y \; +12}$
$\quad 6x \; + y \quad +5$

86.) $\quad 3.2x + 4.3y \; +2$
$\quad \underline{2.7x - 1.1y \; -5}$
$\quad 5.9x + 3.2y \; -3$

81.) $\quad 3j \; -4K \; +6$
$\quad \underline{-4j \; +5K \; -1}$
$\quad -j \; + K \; +5$

82.) $\quad 3xy -4x +y \; -1$
$\quad \underline{2xy +7x \qquad\; +5}$
$\quad 5xy +3x +y \; +4$

87.) $\quad 4.4x - 2.7y$
$\qquad\qquad 3.8y \; -8$
$\quad \underline{2.9x \qquad\quad +3}$
$\quad 7.3x +1.1y \; -5$

83.) $\quad -2m -4n \; +6$
$\quad \underline{3m \; -2n \; -5}$
$\quad +m \; -6n \; +1$

Additional Exercises

Do the following word problems.
1. If an object sells for x dollars then how much will 8 of them cost?

2. If a man has d dollars and then loses x dollars, how much does he have left?

3. If an object costs y dollars, how much will n of them cost?

When y is 3, what does each of the following stand for?
4. $2y$
5. $4y$
6. $9y$
7. $-y$
8. $-5y$
9. $-3y$
10. $y + 4$
11. $7 + y$
12. $y - 2$
13. $y - 9$
14. $3y + 1$
15. $2y - 10$
16. $5 - 4y$

When y is -5, what does each of the following stand for?
17. $2y$
18. $4y$
19. $9y$
20. $-y$
21. $-5y$
22. $-3y$
23. $y + 4$
24. $7 + y$
25. $y - 2$
26. $y - 9$
27. $3y + 1$
28. $2y - 10$
29. $5 - 4y$

When a is 2 and b is 3, what does each of the following stand for?
30. ab
31. $a + b + 1$
32. $2a + 4b$
33. $3a - 2b + 17$

34. $7ab + a$
35. $-2ab + 5$

When x is 3.2 and y is -0.4, what does each of the following stand for?
36. $-3x + y + 2$
37. $2x - 4y - 3$
38. $-5x + 2y$
39. $-4x - 3y$

Do the following word problems.
40. Write an expression that shows the total bill for buying several shirts that cost $6.00 each and several pairs of pants that cost $7.95 each.

$x(6.00) + x(7.95$

41. Use the expression in exercise 40 to find the total bill for buying 3 shirts and 2 pairs of pants.

42. Write an expression that shows the total bill for buying several notebooks that cost $1.95 each and several pens that cost $1.25 each.

43. Use the expression in exercise 42 to find the total bill for buying 5 notebooks and 4 pens.

Add the following monomials.
44. y and y
45. x and $3x$
46. $3a$ and $4a$
47. x and y
48. x and $-y$
49. $2x$ and $5x$
50. $2xy$ and $-4xy$
51. $7x$ and $3x$ and $-4x$
52. $6a$ and $-7a$
53. $-4b$ and $-9b$ and $+9b$

Combine like terms to simplify each of the following polynomials.
54. $4x + 2x$
55. $3x + x + 5x$
56. $7x + 3x - 7x$
57. $2x - 5x - 9x$
58. $4x + x + 3y$
59. $-2x + 4x - 2y - 7 - 2x$
60. $8a + 2ab - 3b + 7a - 5b$
61. $3x - 4y + 7x - 2x - 3y$
62. $3x - 8y + 4x - 7x + 8y$

63. $2x + 4y + 7x + x - 3y$
64. $3x - 5y - 4x - 6y - x$

Add the following polynomials.
65. $x + 2$ and $3x + 4$
66. $2x - 4$ and $5x + 7$
67. $3x + y$ and $2x + 5y$
68. $3x + 4$ and $-4x - 8$
69. $7x + 5$ and $4x + 8$
70. $9x + 2y$ and $-5x + 5y$
71. $-x + y$ and $2y - x$
72. $6b - 5c$ and $2b + 3c$ and $-4b + c$
73. $x + y + z$ and $x - y + z$
74. $5a + 6ab - 7b$ and $-6a - 4b - 2ab$
75. $-3x + 5y$ and $2x - y$ and $4x + 3y$
76. $4.2x + 5.7y$ and $1.3x - 4.2y$
77. $5.8x + 3.4y$ and $-2.7x - 8$
78. $3.6b - 5.1c$ and $-2.7b + 12.7c$
79. $1.5a + 4.9b$ and $4.9a - 3.8b - 5$

Add the following polynomials.
80. $3x + 4y + 5$
     $\underline{2x + y - 2}$

81. $2a - 5b + 3$
     $\underline{7a + 2b - 8}$

82. $3x - 5y - 4z$
     $\underline{-7x - 9y + z}$

83. $-3b - 5c + 8$
     $\underline{2b - 4c - 7}$

84. $3x + 5y + 7xy$
     $- 4y - 9xy$
     $\underline{-5x - 5y}$

85. $8a + 3b - ab + 6$
     $2a - b + ab$
     $+ 2b - 2ab + 7$
     $\underline{6a - b + ab - 2}$

86. $4.7x + 5.8y + 9$
$\underline{3.4x - 2.4y - 6}$

87. $5.1x - 3.2y$
$4.6y + 2$
$\underline{-8.2x - 2}$

Subtraction of Polynomials

To subtract polynomials, we use the same method we used when we subtracted signed numbers. Instead of subtracting, we add the opposite of the second polynomial.

Our first job is to find out what we mean by the opposite of a polynomial. Recall that 9 and -9 are opposites. Note that $9 + (-9) = 0$. In fact, whenever you add a number and its opposite, you get zero.

$2x$ and $-2x$ are opposites, since when you add them, you get zero.

$$2x + (-2x) = 0$$

$(-5x + 3y)$ and $(5x - 3y)$ are opposites since

$$(-5x + 3y) + (5x - 3y) = 0$$

$(3 - 2x + 7y)$ and $(-3 + 2x - 7y)$ are opposites since

$$(3 - 2x + 7y) + (-3 + 2x - 7y) = 0$$

Notice that we get the opposite of each polynomial by changing the sign of <u>every</u> term in the polynomial. So if we are given the polynomial $4y - 5x - 1$, its opposite is $-4y + 5x + 1$.

EXAMPLE 1   Subtract $+12x$ from $-2x$. $2x - 12x$

This is written as $(-2x) - (+12x)$.

The opposite of $+12x$ is $-12x$.

So the subtraction problem: $(-2x) - (+12x)$

becomes the addition problem: $(-2x) + (-12x)$.

But $(-2x) + (-12x)$

is $-14x$

So $(-2x) - (+12x) = -14x$.

EXAMPLE 2   Subtract $-3 + 5x$ from $2x$.

This is written as $(2x) - (-3 + 5x)$.

The opposite of $-3 + 5x$ is $3 - 5x$.

So the subtraction problem: $(2x) - (-3 + 5x)$

becomes the addition problem: $(2x) + (3 - 5x)$.

But $(2x) + (3 - 5x)$

is $2x + 3 - 5x$

or $-3x + 3$

So $(2x) - (-3 + 5x) = -3x + 3$.

EXAMPLE 3   Subtract $5y - 1$ from 3.

This is written as $3 - (5y - 1)$. $3 - (5y - 1)$
                                         $3 + (-5y + 1)$

The opposite of $5y - 1$ is $-5y + 1$

So the subtraction problem: $3 - (5y - 1)$

becomes the addition problem: $3 + (-5y + 1)$

which is $3 - 5y + 1$

or $-5y + 4$

So $3 - (5y - 1) = -5y + 4.$

EXAMPLE 4   From $(a - 2b)$ subtract $(3c - 4d)$

This is written as $(a - 2b) - (3c - 4d)$.

The opposite of $3c - 4d$ is $-3c + 4d$.

So the subtraction problem: $(a - 2b) - (3c - 4d)$

becomes the addition problem: $(a - 2b) + (-3c + 4d)$

which is $a - 2b - 3c + 4d$

So $(a - 2b) - (3c - 4d) = a - 2b - 3c + 4d.$

EXAMPLE 5 Do this problem: $(-6s + 5t - 9n) - (-2s + 3t + 7n)$

Subtraction problem: $(-6s + 5t - 9n) - (-2s + 3t + 7n)$

becomes addition problem: $(-6s + 5t - 9n) + (2s - 3t - 7n)$

which is $-6s + 5t - 9n + 2s - 3t - 7n$

or $-4s + 2t - 16n$

So $(-6s + 5t - 9n) - (-2s + 3t + 7n) = -4s + 2t - 16n.$

EXAMPLE 6   Subtract $5z - 9$ from -14.

$(-14) - (5z - 9)$

becomes $(-14) + (-5z + 9)$

which is $-14 - 5z + 9$

or $-5 - 5z$

So $(-14) - (5z - 9) = -5 - 5z.$

EXAMPLE 7   Subtract $3xy + x - 5y$ from $8xy - 2x - y.$

$$(8xy - 2x - y) - (3xy + x - 5y)$$

becomes $(8xy - 2x - y) + (-3xy - x + 5y)$

which is $8xy - 2x - y - 3xy - x + 5y$

or $5xy - 3x + 4y$

So $(8xy - 2x - y) - (3xy - x - 5y) = 5xy - 3x + 4y$

Here is another way of doing Example 7.

$$\begin{array}{r} 8xy - 2x - y \\ -(3xy + x - 5y) \\ \hline \end{array}$$

becomes
$$\begin{array}{r} 8xy - 2x - y \\ +(-3xy - x + 5y) \\ \hline \end{array}$$
which is $5xy - 3x + 4y$

Be careful: If we use this second method, we must line up like terms. Sometimes we may have to do some rearranging.

EXAMPLE 8   Subtract $(x^3 + 2x^2 - 3x)$ from $(3x^3 + 4x^2 + 7x).$

Note: x^3 is shorthand for $(x)(x)(x)$
$2x^2$ is shorthand for $(2)(x)(x)$
$3x^3$ is shorthand for $(3)(x)(x)(x)$

Watch it! x, x^2 and x^3 are not like terms because they don't have exactly the same letters. So in addition problems you cannot combine them.

So $(3x^3 + 4x^2 + 7x) - (x^3 + 2x^2 - 3x)$

becomes $(3x^3 + 4x^2 + 7x) + (-x^3 - 2x^2 + 3x)$

which is $3x^3 + 4x^2 + 7x - x^3 - 2x^2 + 3x$

or $\quad 2x^3 + 2x^2 + 10x$

So $\quad (3x^3 + 4x^2 + 7x) - (x^3 + 2x^2 - 3x) = 2x^3 + 2x^2 + 10x.$

EXAMPLE 9 From $7x^2y - 3xy^2 + x^2y^2$, subtract $9x^2y + xy^2 - 4x^2y^2$.

Note: $7x^2y$ is shorthand for $(7)(x)(x)(y)$
$-3xy^2$ is shorthand for $(-3)(x)(y)(y)$
x^2y^2 is shorthand for $(x)(x)(y)(y)$

$$(7x^2y - 3xy^2 + x^2y^2) - (9x^2y + xy^2 - 4x^2y^2)$$

becomes $(7x^2y - 3xy^2 + x^2y^2) + (-9x^2y - xy^2 + 4x^2y^2)$

which is $\quad 7x^2y - 3xy^2 + x^2y^2 - 9x^2y - xy^2 + 4x^2y^2$

or $\quad -2x^2y - 4xy^2 + 5x^2y^2$

So $(7x^2y - 3xy^2 + x^2y^2) - (9x^2y + xy^2 - 4x^2y^2) = -2x^2y - 4xy^2 + 5x^2y^2.$

EXAMPLE 10 Subtract $\dfrac{3}{4}x - \dfrac{1}{8}y$ from $\dfrac{1}{4}x + \dfrac{7}{8}y$.

Subtraction problem: $\quad \left(\dfrac{1}{4}x + \dfrac{7}{8}y\right) - \left(\dfrac{3}{4}x - \dfrac{1}{8}y\right)$

becomes addition problem: $\left(\dfrac{1}{4}x + \dfrac{7}{8}y\right) + \left(-\dfrac{3}{4}x + \dfrac{1}{8}y\right)$

which is $\quad \dfrac{1}{4}x + \dfrac{7}{8}y - \dfrac{3}{4}x + \dfrac{1}{8}y$

We combine like terms

$$-\dfrac{2}{4}x + \dfrac{8}{8}y$$

which simplifies to $\quad -\dfrac{1}{2}x + y$.

Exercises

1. Subtract $2x$ from $7x$
2. Subtract $3x$ from $5x$
3. Subtract x^2 from $4x^2$
4. Subtract $-2x$ from $3x$
5. Subtract $-5x$ from $-10x$
6. Subtract $2x^2$ from $-3x^2$
7. Subtract $-9x$ from $-3x$
8. Subtract $5x$ from $2x$
9. Subtract $-2x$ from $3x$
10. Subtract $-10x^3$ from $-11x^3$
11. Subtract $(3x + 5z)$ from $(4x + 7z)$
12. Subtract $(2a + 5b)$ from $(2a - 8b)$
13. Subtract $(5x - 3y)$ from $(2x + 6y)$
14. Subtract -12 from $(2x - 3)$
15. Subtract $(-2x + 3z)$ from 0
16. Subtract 0 from $(-2x + 3z)$
17. Subtract $(5a - 12)$ from -3
18. Subtract $2z - x - y$ from $3y + 2x - 3z$
19. Subtract $(2a^3 - 4a^2 + 5a - 1)$ from $(-a^3 + a^2 + a - 1)$
20. From $3a^2b - ab^2 + 2ab$, subtract $-5a^2b + ab^2 - 4ab$
21. From $8x^3y$ subtract $2x^3 + 8x^3y$
22. From $-5x^2 + 2xy + 3y^2$ subtract $-5x^2 + 2xy + 3y^2$
23. From $-5x^2 + 2xy + 3y^2$ subtract $5x^2 - 2xy - 3y^2$
24. $(3x - 7y - 2z) - (2x + 4y - 3z)$
25. $(4a - 5b + 6c) - (-2a + 3b - 7c)$
26. Subtract $\dfrac{2}{3}x - \dfrac{1}{2}y$ from $\dfrac{5}{3}x - \dfrac{1}{2}y$
27. Subtract $\dfrac{3}{4}a - \dfrac{2}{3}b$ from $\dfrac{3}{4}a - \dfrac{2}{3}b$
28. From $\dfrac{4}{5}x + \dfrac{7}{8}y$ subtract $\dfrac{3}{5}x - \dfrac{1}{8}y$
29. From $\dfrac{1}{4}x + \dfrac{5}{8}y$ subtract $\dfrac{3}{4}x + \dfrac{3}{8}y$

Answers to Exercises

1. $(7x) - (2x)$
$(7x) + (-2x)$
$+5x$

2. $(5x) - (3x)$
$(5x) + (-3x)$
$+2x$

3. $(4x^2) - (x^2)$
$(4x^2) + (-x^2)$
$3x^2$

4. $(3x) - (-2x)$
$(3x) + (+2x)$
$5x$

5. $(-10x) - (-5x)$
$(-10x) + (+5x)$
$-5x$

6. $(-3x^2) - (2x^2)$
$(-3x^2) + (-2x^2)$
$-5x^2$

7. $(-3x) - (-9x)$
$(-3x) + (+9x)$
$6x$

8. $(2x) - (5x)$
$(2x) + (-5x)$
$-3x$

9. $(3x) - (-2x)$
$(3x) + (+2x)$
$5x$

10. $(-11x^3) - (-10x^3)$
$(-11x^3) + (+10x^3)$
$-x^3$

11. $(4x + 7z) - (3x + 5z)$
$(4x + 7z) + (-3x - 5z)$
$4x + 7z - 3x - 5z$
$+x + 2z$

12. $(2a - 8b) - (2a + 5b)$
$(2a - 8b) + (-2a - 5b)$
$2a - 8b - 2a - 5b$
$-13b$

13. $(2x + 6y) - (5x - 3y)$
$(2x + 6y) + (-5x + 3y)$
$2x + 6y - 5x + 3y$
$-3x + 9y$

14. $(2x - 3) - (-12)$
$(2x - 3) + (+12)$
$2x - 3 + 12$
$2x + 9$

(15.) $0-(-2x+3z)$
$0+(+2x-3z)$
$0+2x-3z$
$2x-3z$

(16.) $(-2x+3z)-(0)$
$(-2x+3z)+(0)$
$-2x+3z+0$
$-2x+3z$

(17.) $(-3)-(5a-12)$
$(-3)+(-5a+12)$
$-3-5a+12$
$-5a+9$

(18.) $(3y+2x-3z)-(2z-x-y)$
$(3y+2x-3z)+(-2z+x+y)$
$3y+2x-3z-2z+x+y$
$3x+4y-5z$

(19.) $-a^3+a^2+a-1$
$\underline{-(2a^3-4a^2+5a-1)}$ becomes

$-a^3+a^2+a-1$
$\underline{+(-2a^3+4a^2-5a+1)}$
$-3a^3+5a^2-4a$

(20.) $3a^2b-ab^2+2ab$
$\underline{-(-5a^2b+ab^2-4ab)}$ becomes

$3a^2b-ab^2+2ab$
$\underline{+5a^2b-ab^2+4ab}$
$8a^2b-2ab^2+6ab$

(21.) $8x^3y-(2x^3+8x^3y)$
$8x^3y+(-2x^3-8x^3y)$
$8x^3y-2x^3-8x^3y$
$-2x^3$

(22.)
$-5x^2+2xy+3y^2$
$\underline{-(-5x^2+2xy+3y^2)}$ becomes

$-5x^2+2xy+3y^2$
$\underline{5x^2-2xy-3y^2}$
0

(23.)
$(-5x^2+2xy+3y^2)-(5x^2-2xy-3y^2)$
$(-5x^2+2xy+3y^2)+(-5x^2+2xy+3y^2)$
$-5x^2+2xy+3y^2-5x^2+2xy+3y^2$

$-10x^2+4xy+6y^2$

(24.)
$(3x-7y-2z)-(2x+4y-3z)$
$(3x-7y-2z)+(-2x-4y+3z)$
$3x-7y-2z-2x-4y+3z$
$x-11y+z$

(25.)
$(4a-5b+6c)-(-2a+3b-7c)$
$(4a-5b+6c)+(+2a-3b+7c)$
$4a-5b+6c+2a-3b+7c$
$6a-8b+13c$

26. $\left(\frac{5}{3}x - \frac{1}{2}y\right) - \left(\frac{2}{3}x - \frac{1}{2}y\right)$

$\left(\frac{5}{3}x - \frac{1}{2}y\right) + \left(-\frac{2}{3}x + \frac{1}{2}y\right)$

$\frac{5}{3}x - \frac{1}{2}y - \frac{2}{3}x + \frac{1}{2}y$

$\frac{3}{3}x$ or x

27. $\left(\frac{3}{4}a - \frac{2}{3}b\right) - \left(\frac{3}{4}a - \frac{2}{3}b\right)$

$\left(\frac{3}{4}a - \frac{2}{3}b\right) + \left(-\frac{3}{4}a + \frac{2}{3}b\right)$

$\frac{3}{4}a - \frac{2}{3}b - \frac{3}{4}a + \frac{2}{3}b$

0

28. $\left(\frac{4}{5}x + \frac{7}{8}y\right) - \left(\frac{3}{5}x - \frac{1}{8}y\right)$

$\left(\frac{4}{5}x + \frac{7}{8}y\right) + \left(-\frac{3}{5}x + \frac{1}{8}y\right)$

$\frac{4}{5}x + \frac{7}{8}y - \frac{3}{5}x + \frac{1}{8}y$

$\frac{1}{5}x + y$

29. $\left(\frac{1}{4}x + \frac{5}{8}y\right) - \left(\frac{3}{4}x + \frac{3}{8}y\right)$

$\left(\frac{1}{4}x + \frac{5}{8}y\right) + \left(-\frac{3}{4}x - \frac{3}{8}y\right)$

$\frac{1}{4}x + \frac{5}{8}y - \frac{3}{4}x - \frac{3}{8}y$

$-\frac{2}{4}x + \frac{2}{8}y$ or

$-\frac{1}{2}x + \frac{1}{4}y$

Additional Exercises

1. Subtract $3x$ from $8x$
2. Subtract $5x$ from $7x$
3. Subtract $2x^2$ from $3x^2$
4. Subtract $-6x$ from $2x$
5. Subtract $-4x$ from $-9x$
6. Subtract $4x^2$ from $-7x^2$
7. Subtract $-10x$ from $-x$
8. Subtract $8x$ from $7x$
9. Subtract $-5x$ from $3x$
10. Subtract $-7x^5$ from $-8x^5$
11. Subtract $(2x + z)$ from $(3x + 5z)$
12. Subtract $(3a - b)$ from $(a - 2b)$
13. Subtract $(2x - 2y)$ from $(8x + 5y)$
14. Subtract 7 from $(5 + 3y)$
15. Subtract $(-2a + 7b)$ from 0
16. Subtract 0 from $(-2a + 7b)$
17. Subtract $(8 - 4z)$ from -4
18. Subtract $3a - b - c$ from $4a - 5b - 6c$
19. Subtract $(-3y^3 + 4y^2 - y + 1)$ from $(y^3 - y^2 + y - 1)$
20. From $-ab^2 + ab^3 - ab$, subtract $2ab^2 + ab - 2ab^3$
21. From $2c^3d$ subtract $2cd + 3c^3d$
22. From $7a^2 - 2ay - y^2$ subtract $7a^2 + 2ay + y^2$
23. From $8ab + 9b - 3a$ subtract $-3a - 2b + b^2$
24. $(4x - 8y - 3z) - (3x + 7y - 5z)$
25. $(6a - 8b + 5c) - (-3a + 4b - 9c)$
26. Subtract $\dfrac{4}{5}x - \dfrac{1}{3}y$ from $\dfrac{7}{5}x - \dfrac{2}{3}y$
27. Subtract $\dfrac{1}{2}a - \dfrac{3}{7}b$ from $\dfrac{1}{2}a - \dfrac{3}{7}b$
28. From $\dfrac{3}{4}x + \dfrac{2}{5}y$ subtract $\dfrac{1}{4}x - \dfrac{1}{5}y$
29. From $\dfrac{2}{9}x + \dfrac{3}{8}y$ subtract $\dfrac{4}{9}x + \dfrac{3}{8}y$

Exponents and Multiplication of Monomials

Multiplication of Monomials

When we multiply numbers, the order in which we multiply does not matter. For instance, we can get the answer to $(2)(3)(4)(5)$ in several ways:

$$(2)(3)(4)(5) \qquad (2)(5)(3)(4) \qquad (3)(5)(4)(2)$$

$$6 \quad 20 \qquad\qquad 10 \quad 12 \qquad\qquad 15 \quad 8$$

$$120 \qquad\qquad\quad 120 \qquad\qquad\quad 120$$

In algebra, we can also change the order when we multiply. For example,

$$(4)(x)(5)(y)$$

can be written as

$$(4)(5)(x)(y)$$

$$20 \quad xy$$

$$20xy$$

Notice that in this example we changed the order so that the numbers were all together and the letters were all together.

EXAMPLE 1 Multiply $5x$ by $6y$.

Since $5x$ means $(5)(x)$ and since $6y$ means $(6)(y)$, we can write the problem

$$(5x)(6y) \quad \text{as} \quad (5)(x)(6)(y)$$

Now we can change the order.

$$(5)(6)(x)(y)$$

$$30 \quad xy$$

$$30xy$$

EXAMPLE 2 Multiply $3r$ by $2s$.

$$(3r)(2s)$$

is the same as $(3)(r)(2)(s)$

Since we can change order in multiplication, we can write

$$(3)(2)(r)(s)$$

which is $6rs$

So $(3r)(2s)$ is $6rs$.

EXAMPLE 3 Multiply x by x^2.

$$(x)(x^2)$$

is $(x)(xx)$ (remember x^2 means xx)

which is xxx

which is x^3

So $(x)(x^2)$ is x^3.

EXAMPLE 4   Multiply a^2 by a^3.

$$(a^2)(a^3)$$

is $(aa)(aaa)$

which is $aaaaa$

which is a^5

So $(a^2)(a^3)$ is a^5.

EXAMPLE 5   Multiply 2^2 by 2^3.

$$(2^2)(2^3)$$

which is $(2)(2)(2)(2)(2)$

or 2^5

or 32

So $(2^2)(2^3)$ is 32.

EXAMPLE 6   Multiply $2x^2y^3$ by $4x^2y^4$.

$$(2x^2y^3)(4x^2y^4)$$

which is $(2xxyyy)(4xxyyyy)$

is the same as $(2)(4)xxxxyyyyyyy$

which is $8x^4y^7$

So $(2x^2y^3)(4x^2y^4)$ is $8x^4y^7$.

EXAMPLE 7   Multiply $(5a^2b)$ by $(2a^3b^2)$ by $(4ab^4)$.

$$(5a^2b)(2a^3b^2)(4ab^4)$$

can be written $(5aab)(2aaabb)(4abbbb)$

which is $40a^6b^7$

Note: We multiplied the numbers 5, 2 and 4 to get 40; we counted the number of a's and found there were 6; we counted the number of b's and found there were 7.

So $(5a^2b)(2a^3b^2)(4ab^4)$ is $40a^6b^7$.

EXAMPLE 8   Do each of the following multiplications.

$(4x^2)(-3x^5)$	is the same as	$-12x^7$
$(-5x^2y)(-2x^3y^2)$	is the same as	$10x^5y^3$
$(2b)(3a^2bc)(-4ab^2)$	is the same as	$-24a^3b^4c$
$(a^{23}r^{99})(a^{40}r)$	is the same as	$a^{63}r^{100}$

EXAMPLE 9   A tile is 12.5 centimeters on each side. Find its area.

The expression for the area of a square is s^2 where s is the length of each side. Since s is 12.5,

$$s^2$$

becomes $(12.5)^2$

which is   156.25

So the area of the tile is 156.25 square centimeters.

Exponents

Recall that

x^1 means x
x^2 means xx
x^3 means xxx
x^4 means $xxxx$

Each of the raised numbers is called an **exponent** or **power**.

$3a^2$ means $3aa$
$(3a)^2$ means $(3a)(3a)$ or $9aa$

So $3a^2$ and $(3a)^2$ mean different things.

Be careful: Parentheses can change the meaning of an expression.

Each of the following expressions contains parentheses:

$(-2b)^3$ means $(-2b)(-2b)(-2b)$ or $-8b^3$
$(4x)^2$ means $(4x)(4x)$ or $16x^2$
$(-3y)^4$ means $(-3y)(-3y)(-3y)(-3y)$ or $81y^4$
$(xy)^3$ means $(xy)(xy)(xy)$ or x^3y^3
$3(xy)^2$ means $3(xy)(xy)$ or $3x^2y^2$ ⎱ *Be careful:*
$(3xy)^2$ means $(3xy)(3xy)$ or $9x^2y^2$ ⎰ These are not
$(x^2)^3$ means $(x^2)(x^2)(x^2)$ or x^6 the same.
$(y^3)^4$ means $(y^3)(y^3)(y^3)(y^3)$ or y^{12}
$(2x^2y)^3$ means $(2x^2y)(2x^2y)(2x^2y)$ or $8x^6y^3$ ⎱ *Be careful:*
$2(x^2y)^3$ means $2(x^2y)(x^2y)(x^2y)$ or $2x^6y^3$ ⎰ These are not
   the same.

EXAMPLE 10   Evaluate each of the following expressions.

0^3 means $(0)(0)(0)$ or 0
1^4 means $(1)(1)(1)(1)$ or 1
$(-1)^2$ means $(-1)(-1)$ or 1
$(-1)^3$ means $(-1)(-1)(-1)$ or -1
2^3 means $(2)(2)(2)$ or 8
$(-2)^3$ means $(-2)(-2)(-2)$ or -8
$\left(\dfrac{2}{3}\right)^2$ means $\left(\dfrac{2}{3}\right)\left(\dfrac{2}{3}\right)$ or $\dfrac{4}{9}$
$(0.3)^3$ means $(0.3)(0.3)(0.3)$ or 0.027

Be careful: This next kind is tricky.

-2^3 means $-(2^3)$ or $-(2)(2)(2)$ which is -8
$-x^3$ means $-(x)^3$
-3^2 means $-(3^2)$ or $-(3)(3)$ which is -9
$(-3)^2$ means $(-3)(-3)$ which is 9

Since letters stand for numbers, let's see what $3x^2$ means when x is 5.
When we put 5 in for x, $3x^2$ means $3(5)^2$ or $3(5)(5)$, which is 75.

EXAMPLE 11   What is $(3b)^2$ when b is 2?

Put 2 in for b.

$(3 \cdot 2)^2$

is $(6)^2$

which is 36.

EXAMPLE 12   What is $(3b)^2$ when b is $\dfrac{1}{2}$?

Put $\dfrac{1}{2}$ in for b.

$$\left(3 \cdot \dfrac{1}{2}\right)^2$$

is $\qquad \left(\dfrac{3}{2}\right)^2$

tricky looking

which is $\dfrac{9}{4}$.

EXAMPLE 13   What is $(3b)^2$ when b is -2?

Put -2 in for b.

$$[(3)(-2)]^2$$

or $\qquad (-6)^2$

which is $(-6)(-6)$

or $\qquad 36.$

Note: When we have one set of parentheses inside another, we can use brackets [ ] for one set of parentheses ( ). Thus $((3)(-2))^2$ can be written $[(3)(-2)]^2$. It is important to remember that parentheses (or brackets) can only be used in pairs.

EXAMPLE 14   What is $(2x^2y)^3$ when x is 1 and y is -2?

Put in 1 for x and -2 for y.

$$[2(1)^2(-2)]^3$$

is the same as $\ [2(1)(-2)]^3$

or $\qquad [-4]^3$

which is $\qquad (-4)(-4)(-4)$ or $-64.$

EXAMPLE 15   What is $(2x^2y)^3$ when x is $\dfrac{1}{3}$ and y is $\dfrac{9}{20}$?

We put in $\dfrac{1}{3}$ for x and $\dfrac{9}{20}$ for y.

$$\left[2\left(\frac{1}{3}\right)^2\left(\frac{9}{20}\right)\right]^3$$

which is $\left[2\left(\dfrac{1}{3}\right)\left(\dfrac{1}{3}\right)\left(\dfrac{9}{20}\right)\right]^3$

which is $\left(\dfrac{1}{10}\right)^3$ or $\dfrac{1}{1000}$.

EXAMPLE 16   Add $(2x)^2$ and $(3x)^2$.

$(2x)^2$ means $(2x)(2x)$ which is $4x^2$

and $(3x)^2$ means $(3x)(3x)$ which is $9x^2$

so $(2x)^2 + (3x)^2$

is the same as $(2x)(2x) + (3x)(3x)$

or $4x^2 + 9x^2$

But these are like terms, so we get $13x^2$.

EXAMPLE 17   Add $2x^2$ and $(3x)^2$.

$(3x)^2$ means $(3x)(3x)$ which is $9x^2$

So $2x^2 + (3x)^2$

is the same as $2x^2 + (3x)(3x)$

or $2x^2 + 9x^2$

But these are like terms, so we get $11x^2$.

EXAMPLE 18   Simplify the following expression by combining like terms:

$$3x^3 - (2x)^3 - x^3 + (3x)^3$$

is the same as $3x^3 - (2x)(2x)(2x) - x^3 + (3x)(3x)(3x)$

or $3x^3 - 8x^3 - x^3 + 27x^3$

which is $21x^3$.

Sometimes the terms of a polynomial cannot be combined into one term.

EXAMPLE 19   Simplify the following expression by combining like terms:
$3x^2 + 2x^3 - 4x^3 + 2x^2$

Since $3x^2$ and $2x^2$ are like terms, we can combine them and get $5x^2$.

Since $2x^3$ and $-4x^3$ are like terms, we can combine them and get $-2x^3$.

Since $5x^2$ and $-2x^3$ are unlike terms, this is as far as we can go.

So $3x^2 + 2x^3 - 4x^3 + 2x^2$ can be written as $5x^2 - 2x^3$.

Exercises

Do the following multiplication problems.

1. $(x)(x)$
2. $(x)(x^2)$
3. $(x^2)(x^3)$
4. $(x^4)(x^2)$
5. $(x)(x^5)$
6. $(x^2)(x)(x^3)$
7. $(2x)(3x)$
8. $(2x)(3x^2)$
9. $(5x)(-3x)$
10. $(-4x)(2)(-x)$
11. $(-3)(x)(x^3)$
12. $(x)(y)$
13. $(x)(y^2)$
14. $(xy)(x)$
15. $(xy)(xy)$
16. $(x^2y)(xy^2)$
17. $(-x^3)(xy^4)$
18. $(2x)(-3x)$
19. $(-4x^2)(-2)(3x^3)$
20. $(3x^3)(-2x^4)$
21. $(-2x^3)(-x)(x^5)$
22. $(3b)(4a^2b^2)(-2a^3b^4)$
23. $(x^{70}y^{30})(x^{20}y)(x^{10}y^{10})$
24. $(3^x)(3^y)$
25. $(a^3b^2)(-a^3b^2)$
26. $(2^4)(2)(2^2)$
27. $(2a^5b^4c^7)(-3a^2c^8)(-4bc)$

Do the following word problems.

28. A tile is 15.2 centimeters on each side. Find its area.

29. A rug is 9 ft by 12 ft. Find its area.

30. A room is 12 ft wide and 18 ft long. Find its area.

31. A rug is 4 yd long and 3 yd wide. Find its area.

Write an expression that does not contain parentheses, but means the same thing as each of the following.

32. $(4a)^2$
33. $(ab)^3$
34. $(xy)^2$

35. $(3a)^3$
36. $(2x)^3$
37. $(-2x)^3$
38. $(3a)^2$
39. $(-3a)^2$
40. $-(3a)^2$
41. $(-3x)^3$
42. $5(xy^2)^2$
43. $(5xy^2)^2$
44. $(a^3)^3$
45. $(a^2)^3$
46. $(a^2)(a^3)$
47. $(-2a^2)^3$
48. $(-x^3y)^3$
49. $\left(\dfrac{a^2}{2}\right)^4$
50. $-3(a^2b^3)^3$
51. $(-3a^2b^3)^3$

52. What is $(3x)^2$ when x is 2?
53. What is $3x^2$ when x is 2?
54. What is $(2b)^2$ when b is -2?
55. What is $2b^2$ when b is -2?
56. What is $(2b)^2$ when b is $\dfrac{1}{2}$?
57. What is $2b^2$ when b is $\dfrac{1}{2}$?
58. What is $(3xy^2)^2$ when x is -2 and y is 2?
59. What is $\left(\dfrac{-2ab}{3}\right)^3$ when a is -1 and b is 2?
60. What is $(-ab^2)^3$ when a is 0 and b is 1?
61. Add $(-2a)^3$ and $(3x)^3$
62. Add $3x^3$ and $(-2x)^3$
63. Add $-2x^3$ and $(-2x)^3$
64. Add $(-4x)^2$ and $4(x^2)^2$
65. Combine wherever possible $5x^3 - 3x^3 + (2x)^3$.
66. Combine wherever possible $5y^4 - (3y)^2 + (2y)^4 - 4y^4$.
67. Combine wherever possible $(-2x)^3 + (3x^3)^2 - 4x^3 - x^6$.
68. Combine wherever possible $(7x)^2 - 49x^2$.

Answers to Exercises

(1.) $(x)(x)$
x^2

(2.) $(x)(x^2)$
xxx or x^3

(3.) $(x^2)(x^3)$
$xxxxx$ or x^5

(4.) $(x^4)(x^2)$
$xxxxxx$
x^6

(5.) $(x)(x^5)$
$xxx\,xxx$
x^6

(6.) $(x^2)(x)(x^3)$
$xx\,x\,xx$
x^6

(7.) $(2x)(3x)$
$2\cdot3\cdot x\cdot x$
$6x^2$

(8.) $(2x)(3x^2)$
$2\cdot3\cdot x\cdot x\cdot x$
$6x^3$

(9.) $(5x)(-3x)$
$5\cdot(-3)\cdot x\cdot x$
$-15x^2$

(10.) $(-4x)(2)(-x)$
$(-4)(2)(-1)\,xx$
$8x^2$

(11.) $(-3)(x)(x^3)$
$(-3)\,xxxx$
$-3x^4$

(12.) $(x)(y)$
xy

(13.) $(x)(y^2)$
$x\,y\,y$
$x\,y^2$

(14.) $(xy)(x)$
$x\cdot x\cdot y$
x^2y

(15.) $(xy)(xy)$
$x\cdot x\cdot y\cdot y$
x^2y^2

(16.) $(x^2y)(xy^2)$
$x\cdot x\cdot x\cdot y\cdot y\cdot y$
x^3y^3

⑰ $(-x^3)(xy^4)$
$(-1)\,x\cdot x\cdot x\cdot x\cdot y\cdot y\cdot y\cdot y$
$-x^4 y^4$

⑱ $(2x)(-3x)$
$(2)(-3)\,x\cdot x$
$-6x^2$

⑲ $(-4x^2)(-2)(3x^3)$
$(-4)(-2)(3)\,x\cdot x\cdot x\cdot x\cdot x$
$24x^5$

⑳ $(3x^3)(-2x^4)$
$(3)(-2)\,x\cdot x\cdot x\cdot x\cdot x\cdot x\cdot x$
$-6x^7$

㉑ $(-2x^3)(-x)(x^5)$
$(-1)(-2)\,x\,x\,x\,x\,x\,x\,x\,x\,x$
$2x^9$

㉒ $(3b)(4a^2 b^2)(-2a^3 b^4)$
$3b\cdot 4aa\ bb\ (-2a^3 b^4)$
$-12aa\ bbb\ 2aaa\ bbbb$
$-24a^5 b^7$

㉓ $(x^{70}g^{30})(x^{20}y)(x^{10}y^{10})$
$(x^{90}g^{31})\,(x^{10}y^{10})$
$x^{100}\,y^{41}$

㉔ $(3^x)(3^y)$
$\underbrace{(3)(3)\cdots (3)}_{x\text{ of these}}\ \underbrace{(3)(3)\cdots (3)}_{y\text{ of these}}$
3^{x+y}

㉕ $(a^3 b^2)(-a^3 b^2)$
$-aaaaaa\ bbbb$
$-a^6 b^4$

㉖ $(2^4)\,(2)\,(2^2)$
$(2)(2)(2)(2)(2)(2)(2)$
$2^7\ \text{or}\ 128$

㉗ $(2a^5 b^4 c^7)(-3a^2 c^8)(-4bc)$
$(2)(-3)(-4)\,a^5 a^2 b^4 b\ c^7 c^8 c$
$24\,a^7\,b^5\,c^{16}$

㉘ area of a square
is S^2
Area is $(15.2)^2$
or $231.04\ cm^2$

㉙ area of a rec-
tangle is length ·
width.
Area is $(9)(12)$
or $108\ ft.^2$

30.) area of a rec-
tangle is length ·
width.
area is (12) (18)
or 216 ft.²

31.) area of a rec-
tangle is length ·
width.
area is (4)(3)
or 12 yd.²

32.) $(4a)^2$
$(4a)(4a)$
$16a^2$

33.) $(ab)^3$
$(ab)(ab)(ab)$
$(a^2b^2)(ab)$
a^3b^3

34.) $(xy)^2$
$(xy)(xy)$
x^2y^2

35.) $(3a)^3$
$(3a)(3a)(3a)$
$(9a^2)(3a)$
$27a^3$

36.) $(2x)^3$
$(2x)(2x)(2x)$
$(4x^2)(2x)$
$8x^3$

37.) $(-2x)^3$
$(-2x)(-2x)(-2x)$
$(4x^2)(-2x)$
$-8x^3$

38.) $(3a)^2$
$(3a)(3a)$
$9a^2$

39.) $(-3a)^2$
$(-3a)(-3a)$
$+9a^2$

40.) $-(3a)^2$
$-(3a)(3a)$
$-9a^2$

41.) $(-3x)^3$
$(-3x)(-3x)(-3x)$
$(9x^2)(-3x)$
$-27x^3$

42.) $5(xy^2)^2$
$5(xy^2)(xy^2)$
$5xyy \cdot xyy$
$5x^2y^4$

43. $(5xy^2)^2$
$(5xy^2)(5xy^2)$
$5xyy \ 5xyy$
$25x^2y^4$

44. $(a^3)^3$
$(a^3)(a^3)(a^3)$
$a \cdot a \cdot a \cdot a \cdot a \cdot a \cdot a \cdot a \cdot a$
a^9

45. $(a^2)^3$
$(a^2)(a^2)(a^2)$
$a \cdot a \cdot a \cdot a \cdot a \cdot a$
a^6

46. $(a^2)(a^3)$
$(a \cdot a)(a \cdot a \cdot a)$
$a \cdot a \cdot a \cdot a \cdot a$
a^5

47. $(-2a^2)^3$
$(-2a^2)(-2a^2)(-2a^2)$
$-2 \cdot a \cdot a \cdot (-2) a \cdot a \cdot (-2) a \cdot a$
$-8a^6$

48. $(-x^3y)^3$
$(-x^3y)(-x^3y)(-x^3y)$
$(x^6y^2)(-x^3y)$
$-x^9y^3$

49. $\left(\dfrac{a^2}{2}\right)^4$
$\underbrace{\left(\dfrac{a^2}{2}\right)\left(\dfrac{a^2}{2}\right)}_{\left(\dfrac{a^4}{4}\right)}\underbrace{\left(\dfrac{a^2}{2}\right)\left(\dfrac{a^2}{2}\right)}_{\left(\dfrac{a^4}{4}\right)}$
$\dfrac{a^8}{16}$

50. $-3(a^2b^3)^3$
$-3(a^2b^3)(a^2b^3)(a^2b^3)$
$(-3a^2b^3)(a^4b^6)$
$-3a^6b^9$

51. $(-3a^2b^3)^3$
$(-3a^2b^3)(-3a^2b^3)(-3a^2b^3)$
$(9a^4b^6)(-3a^2b^3)$
$-27a^6b^9$

52. $(3x)^2$
$(3(2))^2$
6^2
36

53. $3x^2$
$3(2)^2$
$3(4)$
12

(54.) $(2b)^2$
$(2(-2))^2$
$(-4)^2$
16

(55.) $2b^2$
$2(-2)^2$
$2(4)$
8

(56.) $(2b)^2$
$(2(\frac{1}{2}))^2$
1^2
1

(57.) $2b^2$
$2(\frac{1}{2})^2$
$2(\frac{1}{4})$
$\frac{1}{2}$

(58.) $(3xy^2)^2 = (3(-2)(2)^2)^2$
$(3(-2)(4))^2$
$(3(-8))^2$
$(-24)^2$
576

(59.) $\left(-\frac{2ab}{3}\right)^3 = \left(\frac{-2(-1)(2)}{3}\right)^3$
$\left(\frac{(-2)(-2)}{3}\right)^3$
$\left(\frac{4}{3}\right)^3$
$\left(\frac{4}{3}\right)\left(\frac{4}{3}\right)\left(\frac{4}{3}\right)$
$\left(\frac{16}{9}\right)\left(\frac{4}{3}\right) = \frac{64}{27}$

(60.) $(-ab^2)^3$
$(-(0)(1)^2)^3$
$(-(0)(1))^3$
$(-0)^3 = (0)(0)(0)$
0

(61.) $(-2a)^3 + (3x)^3$
$(-2a)(-2a)(-2a) + (3x)(3x)(3x)$
$(4a^2)(-2a) + (9x^2)(3x)$
$-8a^3 + 27x^3$

(62.) $3x^3 + (-2x)^3$
$3x^3 + (-2x)(-2x)(-2x)$
$3x^3 + (4x^2)(-2x)$
$3x^3 + (-8x^3)$
$-5x^3$

(63.) $-2x^3 + (-2x)^3$
$\quad -2x^3 + (-2x)(-2x)(-2x)$
$\quad -2x^3 + (4x^2)(-2x)$
$\quad -2x^3 + (-8x^3)$
$\quad\quad -10x^3$

(68.) $(7x)^2 - 49x^2$
$\quad (7x)(7x) - 49x^2$
$\quad\quad 49x^2 - 49x^2$
$\quad\quad\quad 0$

(64.) $(-4x)^2 + 4(x^2)^2$
$(-4x)(-4x) + 4(x^2)(x^2)$
$\quad 16x^2 + 4x^4$

(65.) $5x^3 - 3x^3 + (2x)^3$
$\quad 2x^3 + (2x)(2x)(2x)$
$\quad 2x^3 + (4x^2)(2x)$
$\quad\quad 2x^3 + 8x^3$
$\quad\quad\quad 10x^3$

(66.) $5y^4 - (3y^2)^2 + (2y)^4 4y^4$
$\quad 5y^4 - 9y^4 + 16y^4 - 4y^4$
$\quad -4y^4 + 12y^4$
$\quad\quad 8y^4$

(67.) $(-2x)^3 + (3x^3)^2 4x^3 - x^6$
$\quad -8x^3 + 9x^6 - 4x^3 - x^6$
$\quad\quad 8x^6 - 12x^3$

Additional Exercises

Do the following multiplication problems.

1. $(a)(a)$
2. $(a^2)(a)$
3. $(a^3)(a^2)$
4. $(a^4)(a^2)$
5. $(a)(a^6)$
6. $(a^3)(a)(a^2)$
7. $(3a)(4a)$
8. $(2a)(-7a)$
9. $(-5a)(3)(-a)$
10. $(-2)(a)(a^4)$
11. $(a)(b)$
12. $(a^2)(b)$
13. $(ab)(a)$
14. $(ab)(ab)$
15. $(a^2b)(ab^2)$
16. $(-a^5)(ab^3)$
17. $(6a)(-3a)$
18. $(-2a^3)(-3)(a^4)$
19. $(-5a^4)(-4a^5)$
20. $(-2a^2)(2a)(2a)$
21. $(8a^4)(-2a^3)(-a)$
22. $3a(-2a)(-1)$
23. $(c^{20}d^{10}f^5)(c^{10}d^5f)(c^5d^{10}f^5)$
24. $(a^2)(a^y)(a)$
25. $(-2)(a^2)(a^5)$
26. $(abc^2)(a^2bc)(ab^2c^2)$
27. $(-2x^{14})(-3)(x^6)$

Do the following word problems.

28. A tile is 14.4 centimeters on each side. Find its area.

29. A rug is 8 ft by 11 ft. Find its area.

30. A blackboard is 6 ft wide and 4 ft long. Find its area.

31. A room is 5 yd long and 3 yd wide. Find its area.

Write an expression that does not contain parentheses, but means the same thing as each of the following.

32. $(5b)^2$
33. $(bc)^2$
34. $(rs)^3$

35. $(4y)^3$
36. $(2x)^4$
37. $(-2x)^4$
38. $(7y)^2$
39. $(-7y)^2$
40. $-(7y)^2$
41. $(-2x)^5$
42. $3(a^2b)^2$
43. $(-4x^2y^3)^2$
44. $(-b^3)^2$
45. $(-b^2)^3$
46. $(-b^2)(b^3)$
47. $(3a^2)^3$
48. $(x^5y^7)^3$
49. $\left(\dfrac{z^2}{3}\right)^2$
50. $-2(x^2y)^3$
51. $(-2x^2y)^3$

52. What is $(4x)^2$ when x is 3?
53. What is $4x^2$ when x is 3?
54. What is $(-3b)^3$ when b is 2?
55. What is $-3b^3$ when b is 2?
56. What is $(4b)^2$ when b is $\dfrac{1}{2}$?
57. What is $4b^2$ when b is $\dfrac{1}{2}$?
58. What is $(2c^2d)^2$ when c is 2 and d is 3?
59. What is $\left(\dfrac{-xy}{2}\right)^3$ when x is 1 and y is -1?
60. What is $\dfrac{xy}{-2}$ when x is 1 and y is 0?
61. Add $5a^2$ and $(-2a)^2$
62. Add $2x^3$ and $(-3x)^3$
63. Add $4x^4$ and $(3x^2)^2$
64. Add $(-x)^3$ and $-2x^3$
65. Combine wherever possible $2a^3 - 4a^3 + (-2a)^3$.
66. Combine wherever possible $(10b)^2 - 10b^2$.
67. Combine wherever possible $(9ac)^2 - 9(ac)^2$.
68. Combine wherever possible $9ac^2 + (9ac)^2 - 9a^2c^2 + a(3c)^2$.

Lesson 5

Division of Monomials by Monomials

A division problem can be written as a fraction.

EXAMPLE 1 Divide 15 by 3.

To divide 15 by 3 we write $\dfrac{15}{3}$. Since 15 split into its primes is (3)(5),

we can write $\dfrac{15}{3} = \dfrac{(3)(5)}{(3)}$

but $\dfrac{3}{3}$ is $\dfrac{1}{1}$ or 1

So $\left(\dfrac{3}{3}\right)5$

is the same as (1)5 or 5

So $\dfrac{15}{3} = 5.$

EXAMPLE 2 Divide 42 by 30.

To divide 42 by 30 we write $\dfrac{42}{30}$

42 split into its primes is (2)(3)(7)
30 split into its primes is (2)(3)(5)

So $\dfrac{42}{30}$ can be written as $\dfrac{(2)(3)(7)}{(2)(3)(5)}$.

We know that $\dfrac{2}{2}$ is $\dfrac{1}{1}$ or 1 and that $\dfrac{3}{3}$ is $\dfrac{1}{1}$ or 1. We will use a slash

when we replace something like $\dfrac{2}{2}$ or $\dfrac{3}{3}$ with 1. We call this **canceling**.

So from now on, when we use a slash we will think of it as $\dfrac{1}{1}$ or 1.

Since $\dfrac{42}{30}$ can be written as $\dfrac{(2)(3)(7)}{(2)(3)(5)}$ we can cancel the 2's and the 3's.

$\dfrac{(\cancel{2})(\cancel{3})(7)}{(\cancel{2})(\cancel{3})(5)}$ is $(1)(1)\,\dfrac{7}{5}$ or $\dfrac{7}{5}$

So $\dfrac{42}{30}$ is $\dfrac{7}{5}$.

EXAMPLE 3 Divide x^3 by x.

To divide x^3 by x, we write $\dfrac{x^3}{x}$.

But x^3

means xxx (This is just like splitting a number into its primes.)

So $\dfrac{x^3}{x}$

is the same as $\dfrac{xxx}{x}$

But $\dfrac{x}{x} = 1$

So $\quad\dfrac{\overset{1}{\cancel{x}}xx}{\underset{1}{\cancel{x}}}$

is the same as xx

or $\quad x^2$

So $\quad \dfrac{x^3}{x} = x^2.$

Let's show that $\dfrac{x^3}{x} = x^2$ is <u>true</u> when x is 2. When x is 2,

then $\qquad \dfrac{x^3}{x} = x^2$

becomes $\quad \dfrac{2^3}{2} = 2^2$

or $\qquad \dfrac{8}{2} = 4 \qquad$ <u>True</u>

Let's show that $\dfrac{x^3}{x} = x^2$ is <u>true</u> when x is -3. When x is -3,

then $\qquad \dfrac{x^3}{x} = x^2$

becomes $\quad \dfrac{(-3)^3}{-3} = (-3)^2 \quad\longleftarrow$

or $\qquad \dfrac{-27}{-3} = 9 \qquad$ <u>True</u>

In fact $\dfrac{x^3}{x} = x^2$ is true no matter what number (except zero) is put in for each x. We cannot put zero in for x since that would give zero in the bottom of the fraction. Remember, division by zero has no meaning.

EXAMPLE 4 Divide y^5 by y^2.

$$\dfrac{y^5}{y^2}$$

is the same as $\dfrac{yyyyy}{yy}$

but $\qquad \dfrac{yy}{yy}$ is $\dfrac{1}{1}$ or 1

So
$$\frac{\cancel{y}\cancel{y}yyy}{\cancel{y}\cancel{y}}$$

is $(1)(1)(yyy)$

which is yyy or y^3

So $\dfrac{y^5}{y^2} = y^3.$

EXAMPLE 5   What is $\dfrac{3x^2y^3z^4}{6xyz^2}$?

$$\frac{3x^2y^3z^4}{6xyz^2}$$

is the same as $\dfrac{\cancel{3}x\cancel{x}\cancel{y}yy\cancel{z}\cancel{z}zz}{(2)(\cancel{3})\cancel{x}\cancel{y}\cancel{z}\cancel{z}}$

which is $\dfrac{xy^2z^2}{2}$

So $\dfrac{3x^2y^3z^4}{6xyz^2} = \dfrac{xy^2z^2}{2}$.

EXAMPLE 6   What is $\dfrac{xy^2z}{x^2y^2}$?

$$\frac{xy^2z}{x^2y^2}$$

is the same as $\dfrac{\cancel{x}\cancel{y}\cancel{y}z}{x\cancel{x}\cancel{y}\cancel{y}}$

or $\dfrac{z}{x}$

So $\dfrac{xy^2z}{x^2y^2} = \dfrac{z}{x}$.

EXAMPLE 7   What is $\dfrac{x^3}{x^3}$?

$$\frac{x^3}{x^3}$$

is the same as $\dfrac{\overset{1\ 1\ 1}{\cancel{x}\cancel{x}\cancel{x}}}{\underset{1\ 1\ 1}{\cancel{x}\cancel{x}\cancel{x}}}$

which is $(1)(1)(1)$

or 1

This makes sense, since anything (except zero) divided by itself is 1.

So $\dfrac{x^3}{x^3} = 1$.

✳ EXAMPLE 8   Divide x^2 by x^5.

$$\frac{x^2}{x^5}$$

is the same as $\dfrac{xx}{xxxxx}$

but $\dfrac{xx}{xx}$ is $(1)(1)$ or 1

so $\dfrac{\overset{1\ 1}{\cancel{x}\cancel{x}}}{\underset{1\ 1}{\cancel{x}\cancel{x}xxx}}$

which is $\dfrac{(1)(1)}{xxx}$

or $\dfrac{1}{xxx}$ which is $\dfrac{1}{x^3}$

So x^2 divided by x^5 is $\dfrac{1}{x^3}$.

⁂**EXAMPLE 9**   Simplify $\dfrac{x^{32}}{x^6}$.

$\dfrac{x^{32}}{x^6}$ is the same as $\dfrac{x^{26}\overset{1}{\cancel{x^6}}}{\underset{1}{\cancel{x^6}}}$

or x^{26}

So $\dfrac{x^{32}}{x^6} = x^{26}$.

EXAMPLE 10   Simplify $\dfrac{y^7}{y}$.

$\dfrac{y^7}{y}$ is the same as $\dfrac{y^6\overset{1}{\cancel{y}}}{\underset{1}{\cancel{y}}}$

or y^6

So $\dfrac{y^7}{y} = y^6$.

⁂**EXAMPLE 11**   Simplify $\dfrac{z^2}{z^5}$.

$\dfrac{z^2}{z^5}$ is the same as $\dfrac{\overset{1}{\cancel{z^2}}}{\underset{1}{\cancel{z^2}}z^3}$

which is $\dfrac{1}{z^3}$

So $\dfrac{z^2}{z^5} = \dfrac{1}{z^3}$.

EXAMPLE 12   Simplify $\dfrac{x^{40}y^{30}}{x^5y^{10}}$.

$\dfrac{x^{40}y^{30}}{x^5y^{10}}$ is the same as $\dfrac{x^{35}\overset{1}{\cancel{x^5}}y^{10}\overset{1}{\cancel{y^{10}}}y^{20}}{\underset{1}{\cancel{x^5}}\underset{1}{\cancel{y^{10}}}}$

or $x^{35}y^{20}$

So $\dfrac{x^{40}y^{30}}{x^5y^{10}} = x^{35}y^{20}.$

EXAMPLE 13   Simplify $\dfrac{8b^{10}a^4}{2b^2a^7}$.

$$\dfrac{8b^{10}a^4}{2b^2a^7} \text{ is the same as } \dfrac{\overset{1}{4}(\overset{1}{\cancel{2}})b^8\overset{1}{\cancel{b^2}}\cancel{a^4}}{\cancel{2}\underset{1}{\cancel{b^2}}\cancel{a^4}\underset{1}{a^3}}$$

which is $\dfrac{4b^8}{a^3}$

So $\dfrac{8b^{10}a^4}{2b^2a^7} = \dfrac{4b^8}{a^3}$.

EXAMPLE 14   Simplify $\dfrac{2a(5b)^2}{5abc}$.

$$\dfrac{2a(5b)^2}{5abc} \text{ is the same as } \dfrac{2\overset{1}{\cancel{a}}(\overset{1}{\cancel{5}}\overset{1}{\cancel{b}})(5b)}{\underset{1}{\cancel{5}}\underset{1}{\cancel{a}}\underset{1}{\cancel{b}}c}$$

which is $\dfrac{(2)(5)(b)}{c}$ or $\dfrac{10b}{c}$

So $\dfrac{2a(5b)^2}{5abc} = \dfrac{10b}{c}$.

EXAMPLE 15   Simplify $\dfrac{(3a)^2(2b)}{ab}$.

$$\dfrac{(3a)^2(2b)}{ab} \text{ is the same as } \dfrac{(3\overset{1}{\cancel{a}})(3a)(2\overset{1}{\cancel{b}})}{\underset{1}{\cancel{a}}\underset{1}{\cancel{b}}}$$

which is $\dfrac{(3)(a)(3)(2)}{1}$ or $18a$

So $\dfrac{(3a)^2(2b)}{ab} = 18a.$

Exercises

Give the answers to each of the following problems. Cancel whenever possible and put the answer in its simplest form.

1. $\dfrac{35}{7}$

2. $\dfrac{46}{30}$

3. $\dfrac{60}{48}$

4. $\dfrac{x^2}{x}$

5. Show that $\dfrac{x^2}{x} = x$ when x is 3.

6. Show that $\dfrac{x^2}{x} = x$ when x is -5.

7. Divide y^4 by y^2.

For each of the following problems do the indicated division.

8. $\dfrac{x^5}{x^3}$

9. $\dfrac{b^4}{b^3}$

10. $\dfrac{4x^5}{2x^2}$

11. $\dfrac{x^5y^2}{x^4y}$

12. $\dfrac{a^4b^3c^2}{a^2b^3c}$

13. $\dfrac{15x^2y^4z}{25xy^2}$

14. $\dfrac{y^2}{y^4}$

15. $\dfrac{a^4b^2}{a^3b^3}$

16. $\dfrac{4a^3b^4c^2}{8a^3b^5c}$

17. $\dfrac{x^2 y^2}{xy}$

18. $\dfrac{a^4 b}{a^2 b^4}$

19. $\dfrac{y^2}{y^6}$

20. $\dfrac{a^4 b^2}{a^3 b^3 c}$

21. $\dfrac{3a^4 b^2 c}{12a^3 b^2 c^3}$

22. $\dfrac{x^3 y}{x^4 y}$

23. $\dfrac{a^4 b^{14}}{a^2 b^3}$

24. $\dfrac{r^{33} s^{22} t^{14}}{r^2 s^4 t^{20}}$

25. If a is 2 and b is -1, what does $\dfrac{a^2 b^3}{ab}$ stand for?

Simplify.

26. $\dfrac{x^{25} y^{40}}{x^5 y^{20}}$

27. $\dfrac{a^{50} b^{20}}{a^{10} b^5}$

Answers to Exercises

① $\frac{35}{7} = \frac{5 \cdot \cancel{7}}{\cancel{7}}$ or 5

② $\frac{46}{30} = \frac{\cancel{2} \cdot 23}{\cancel{2} \cdot 15}$ or $\frac{23}{15}$

③ $\frac{60}{48} = \frac{\cancel{4} \cdot 3 \cdot 5}{\cancel{4} \cdot 3 \cdot 4}$ or $\frac{5}{4}$

④ $\frac{x^2}{x} = \frac{\cancel{x} \cdot x}{\cancel{x}}$ or x

⑤ $\frac{x^2}{x} \;?\; x$

$\frac{3^2}{3} \;?\; 3$

$\frac{9}{3} \;?\; 3$

$3 = 3$

⑥ $\frac{x^2}{x} \;?\; x$

$\frac{(-5)^2}{-5} \;?\; -5$

$\frac{25}{-5} \;?\; -5$

$-5 = -5$

⑦ $\frac{y^4}{y^2} = \frac{\cancel{y}\cancel{y}yy}{\cancel{y}\cancel{y}}$ or y^2

⑧ $\frac{x^5}{x^3} = \frac{\cancel{x}\cancel{x}\cancel{x}xx}{\cancel{x}\cancel{x}\cancel{x}}$ or x^2

⑨ $\frac{b^4}{b^3} = \frac{\cancel{b}\cancel{b}\cancel{b}b}{\cancel{b}\cancel{b}\cancel{b}}$ or b

⑩ $\frac{4x^5}{2x^2} = \frac{\cancel{2} \cdot 2 \cdot \cancel{x} \cdot \cancel{x} \cdot x \cdot x \cdot x}{\cancel{2}\cancel{x}\cancel{x}}$

$2x^3$

⑪ $\frac{x^5 y^2}{x^4 y} = \frac{\cancel{x}\cancel{x}\cancel{x}\cancel{x}x \; \cancel{y}y}{\cancel{x}\cancel{x}\cancel{x}\cancel{x} \; \cancel{y}}$

xy

⑫ $\frac{a^4 b^3 c^2}{a^2 b^3 c} =$

$\frac{\cancel{a}\cancel{a}aa \; \cancel{b}\cancel{b}\cancel{b} \; \cancel{c}c}{\cancel{a}\cancel{a} \; \cancel{b}\cancel{b}\cancel{b} \; \cancel{c}}$

$a^2 c$

⑬ $\frac{15x^2 y^4 z}{25 x y^2} =$

$\frac{3 \cdot 5 x x \; \cancel{y}\cancel{y}yy z}{\cancel{5} \cdot 5 \; \cancel{x} \; \cancel{y}\cancel{y}}$

$\frac{3x y^2 z}{5}$

(14.) $\dfrac{y^2}{y^4} = \dfrac{\cancel{y}\cancel{y}}{\cancel{y}\cancel{y}yy}$ or $\dfrac{1}{y^2}$

(20.) $\dfrac{a^4 b^2}{a^3 b^3 c} =$

$\dfrac{\cancel{a}\cancel{a}\cancel{a}a \; \cancel{b}\cancel{b}}{\cancel{a}\cancel{a}\cancel{a} \; \cancel{b}\cancel{b}b \; c}$

$\dfrac{a}{bc}$

(15.) $\dfrac{a^4 b^2}{a^3 b^3} = \dfrac{\cancel{a}\cancel{a}\cancel{a}a \; \cancel{b}\cancel{b}}{\cancel{a}\cancel{a}\cancel{a} \; \cancel{b}\cancel{b}b}$

$\dfrac{a}{b}$

(16.) $\dfrac{4 a^3 b^4 c^2}{8 a^3 b^5 c} =$

$\dfrac{\cancel{4}\cancel{a}\cancel{a}\cancel{a} \; \cancel{b}\cancel{b}\cancel{b}\cancel{b} \; cc}{2 \cdot \cancel{4}\cancel{a}\cancel{a}\cancel{a} \; \cancel{b}\cancel{b}\cancel{b}\cancel{b}b \; \cancel{c}}$

$\dfrac{c}{2b}$

(21.) $\dfrac{3 a^4 b^2 c}{12 a^3 b^2 c^3} =$

$\dfrac{\cancel{3}\cancel{a}\cancel{a}\cancel{a}a \; \cancel{b}\cancel{b} \; \cancel{c}}{3 \cdot 4 \cancel{a}\cancel{a}\cancel{a} \; \cancel{b}\cancel{b} \cancel{c}cc}$

$\dfrac{a}{4c^2}$

(17.) $\dfrac{x^2 y^2}{x y} = \dfrac{\cancel{x}x \; \cancel{y}y}{\cancel{x} \; \cancel{y}}$

xy

(22.) $\dfrac{x^3 y}{x^4 y} = \dfrac{\cancel{x}\cancel{x}\cancel{x}\cancel{y}}{\cancel{x}\cancel{x}\cancel{x}x \cancel{y}}$

$\dfrac{1}{x}$

(18.) $\dfrac{a^4 b}{a^2 b^4} = \dfrac{\cancel{a}\cancel{a}aa \; \cancel{b}}{\cancel{a}\cancel{a} \; \cancel{b}bbb}$

$\dfrac{a^2}{b^3}$

(23.) $\dfrac{a^4 b^{14}}{a^2 b^3}$

$\dfrac{a^2 \cancel{a}^2 \; \cancel{b}^3 b^{11}}{\cancel{a}^2 \; \cancel{b}^3}$

$a^2 b^{11}$

(19.) $\dfrac{y^2}{y^6} = \dfrac{\cancel{y}\cancel{y}}{\cancel{y}\cancel{y}yyyy}$

$\dfrac{1}{y^4}$

24. $\dfrac{r^{33} s^{22} t^{14}}{r^2 s^4 t^{20}} =$

$\dfrac{r^{31} r^{2} s^{18} s^{4} t^{14} t^{14}}{r^2 s^4 t^{14} t^6}$

$\dfrac{r^{31} s^{18}}{t^6}$

25. $\dfrac{a^2 b^3}{a b} =$

$\dfrac{2^2 (-1)^3}{2(-1)}$

$\dfrac{2 \cdot 2 (-1)}{2(-1)}$

2

26. $\dfrac{x^{25} y^{40}}{x^5 y^{20}} =$

$\dfrac{x^{20} x^{5} y^{20} y^{20}}{x^5 y^{20}}$

$x^{20} y^{20}$

27. $\dfrac{a^{50} b^{20}}{a^{10} b^5} =$

$\dfrac{a^{40} a^{10} b^{15} b^{5}}{a^{10} b^5}$

$a^{40} b^{15}$

Additional Exercises

Give the answers to each of the following problems. Cancel whenever possible and put the answer in its simplest form.

1. $\dfrac{12}{4}$

2. $\dfrac{45}{36}$

3. $\dfrac{70}{42}$

4. $\dfrac{y^3}{y}$

5. Show that $\dfrac{y^3}{y} = y^2$ when y is 4.

6. Show that $\dfrac{y^3}{y} = y^2$ when y is -2.

7. $\dfrac{x^4}{x^2}$

For each of the following problems, do the indicated division.

8. $\dfrac{x^4}{x^3}$

9. $\dfrac{y^5}{y^2}$

10. $\dfrac{6x^4}{3x}$

11. $\dfrac{a^4 b^2}{a^3 b}$

12. $\dfrac{a^6 b^3 c^4}{a^4 b^3 c}$

13. $\dfrac{18x^4 y z^4}{24x^2 z^3}$

14. $\dfrac{y^3}{y^5}$

15. $\dfrac{a^2 b^3}{a\ b^4}$

16. $\dfrac{5x^2 y^3 z^2}{10xy^3 z^4}$

17. $\dfrac{x^4 y^3}{x y^2}$

18. $\dfrac{a^5 b^5}{a^4 b}$

19. $\dfrac{x^3}{x^4}$

20. $\dfrac{a^2 b^5}{a^3 b^3}$

21. $\dfrac{6 a^4 b^2 c^3}{12 a^4 b^4 c}$

22. $\dfrac{a y^3}{a^2 b^2}$

23. $\dfrac{a^5 b^2}{a b^2}$

24. $\dfrac{r^{24} s^{11} t^{12}}{r^3 s^3 t^{20}}$

25. If a is 3 and b is -2, what does $\dfrac{a^3 b^2}{a^2 b}$ stand for?

Simplify.

26. $\dfrac{x^{55} y^{25}}{x^{10} y^{10}}$

27. $\dfrac{a^{30} b^{40}}{a^5 b^{15}}$

Lesson 6

Multiplication of Polynomials

We use the symbol 5( ) to mean 5 times whatever is inside the parentheses.

So $5(3)$ means 5 times 3 or 15
$5(3 + 4)$ means 5 times 7 or 35
$5(3 - 4)$ means 5 times -1 or -5
$-5(3 + 4)$ means -5 times 7 or -35

Let's look at $3(2 + 5)$ and $3(2) + 3(5)$. ✳ Wow

$$3(2 + 5) \quad \| \quad 3(2) + 3(5)$$
$$3(7) \quad \| \quad +6 + 15$$
$$21 \quad \| \quad 21$$

Note that $3(2 + 5)$ is the same as $3(2) + 3(5)$, since they both equal 21.

EXAMPLE 1 Let's look at more examples of the same kind.

$$2(3 + 4) \quad \| \quad 2(3) + 2(4)$$
$$2(7) \quad \| \quad 6 + 8$$
$$14 \quad \| \quad 14$$

91

$$3(-2 + 3) \quad \| \quad 3(-2) + 3(3)$$
$$3(1) \qquad\quad \| \quad -6 + 9$$
$$3 \qquad\qquad \| \quad 3$$

$$3(4 - 2) \quad\; \| \quad 3(4) + 3(-2)$$
$$3(2) \qquad\quad \| \quad 12 - 6$$
$$6 \qquad\qquad \| \quad 6$$

So $2(3 + 4)$ is the same as $2(3) + 2(4)$
      $3(-2 + 3)$ is the same as $3(-2) + 3(3)$
      $3(4 - 2)$ is the same as $3(4) + 3(-2)$

Let's look again at $2(3 + 4)$. *Think:* Multiply each number inside the parentheses by 2.

$$2(3 + 4)$$

is the same as $2(3) + 2(4)$

which is $6 + 8$

or   14.

EXAMPLE 2   Write $5(x + y)$ without using parentheses. $= 5x + 5y$

$$5(x + y)$$

is the same as $5(x) + 5(y)$

which is $5x + 5y$.

EXAMPLE 3   Write $2(3x + 5y)$ without using parentheses.

$$2(3x + 5y)$$

is the same as $2(3x) + 2(5y)$

which is $6x + 10y$.

EXAMPLE 4   Write $4x(1 + 6y)$ without using parentheses.

$$4x(1 + 6y)$$

is the same as $4x(1) + 4x(6y)$

which is $4x + 24xy$.

Since we can change the order in multiplication,

$$(1 + 6y)4x$$

is the same as $4x(1 + 6y)$

So $(1 + 6y)4x$ is the same as $4x + 24xy$.

EXAMPLE 5   Write $(a^3 - b^3)ab^3$ without using parentheses.

$$(a^3 - b^3)ab^3$$

is the same as $a^4b^3 - ab^6$.

EXAMPLE 6   Write $(a + b)c + (a + b)d$ without using parentheses.

$$(a + b)c + (a + b)d$$

is the same as $ac + bc + ad + bd$.

EXAMPLE 7   Write $(a + b)c + c(a + b)$ without using parentheses.

$$(a + b)c + c(a + b)$$

is the same as $ac + bc + ca + cb$

But ca can be written as ac; and cb can be written as bc.

So $ac + bc + ca + cb$ is the same as $ac + bc + ac + bc$

which is $2ac + 2bc$.

EXAMPLE 8   Evaluate $4(5 + 2 + 1)$ by two different methods.

$4(5 + 2 + 1)$	$4(5 + 2 + 1)$
$4(8)$	$4(5) + 4(2) + 4(1)$
32	$20 + 8 + 4$
	32

EXAMPLE 9   Write $a(b + c + d)$ without using parentheses.

$$a(b + c + d)$$

is the same as $ab + ac + ad$.

EXAMPLE 10   Write $3x^2y(2x - 3y + 4)$ without using parentheses.

$$3x^2y(2x - 3y + 4)$$

is the same as $6x^3y - 9x^2y^2 + 12x^2y$.

EXAMPLE 11   Write $-2xyz(3x - 2yz + z^2)$ without using parentheses.

$$-2xyz(3x - 2yz + z^2)$$

is the same as $-6x^2yz + 4xy^2z^2 - 2xyz^3$.

We know that $x(y + z)$ can be written in another form. That is,

$x(y + z)$ is the same as $xy + xz$

Similarly, no matter what the parentheses (   ) contain

($)(y + z)$ is the same as ($)y + ($ $)z$

Let's use this method to multiply $x + 2$ by $x + 3$.

$(x + 2)(x + 3)$ is the same as $(x + 2)x + (x + 2)3$

but $(x + 2)x + (x + 2)3$ is the same as $x^2 + 2x + 3x + 6$

or $x^2 + 5x + 6$

So $(x + 2)(x + 3)$ is $x^2 + 5x + 6$.

EXAMPLE 12   Multiply $(x + 5)$ by $(2x - 3)$.

$(x + 5)(2x - 3)$ is the same as $(x + 5)2x + (x + 5)(-3)$

but $(x + 5)2x + (x + 5)(-3)$ is the same as $2x^2 + 10x - 3x - 15$

or $2x^2 + 7x - 15$

So $(x + 5)(2x - 3)$ is $2x^2 + 7x - 15$.

EXAMPLE 13   Let's do example 12 in a different way. As before, we can work from either side.

$(x + 5)(2x - 3)$ is the same as $x(2x - 3) + 5(2x - 3)$

$x(2x - 3) + 5(2x - 3)$ is the same as $2x^2 - 3x + 10x - 15$

$$2x^2 \qquad 7x \qquad -15$$

or $2x^2 + 7x - 15$.

Note: It makes no difference which side we work from; the answer is the same.

EXAMPLE 14   Write an expression without parentheses which means the same as $(x - 5)^2$.

$$(x-5)(x-5) = x(x-5) +$$
$$-5(x-5)$$
$$x^2 - 5x - 5x + 25$$
$$x^2 - 25x + 25$$
$$x^2 - 10x + 25$$

$(x - 5)^2$ means $(x - 5)(x - 5)$

$(x - 5)(x - 5)$

$(x - 5)x + (x - 5)(-5)$

$x^2 - 5x - 5x + 25$

$x^2 - 10x + 25$

So $(x - 5)^2$ is $x^2 - 10x + 25$.

EXAMPLE 15   Multiply $(x + 5)$ by $(x - 5)$.

$(x + 5)(x - 5)$

$(x + 5)x + (x + 5)(-5)$

$x^2 + 5x - 5x - 25$

$x^2 - 25$

So $(x + 5)(x - 5)$ is $x^2 - 25$.

We can multiply polynomials in another way. Let's multiply $(x + 5)$ by $(2x - 3)$.

$$\begin{array}{r} x + 5 \\ 2x - 3 \\ \hline 2x^2 \quad 2 \end{array}$$

Here we multiply $(x + 5)$ by each term of $(2x - 3)$. That is, first multiply $(x + 5)$ by $2x$ and then multiply $(x + 5)$ by -3.

$$
\begin{array}{l}
x \;\; + 5 \\
2x \;\; - 3 \\
\hline
2x^2 + 10x \qquad \text{This line is } 2x(x + 5) \\
\;\;\;\; - 3x - 15 \qquad \text{This line is } -3(x + 5) \\
\hline
2x^2 + \;\; 7x - 15
\end{array}
$$

add

As you work, be sure to put like terms under each other.

EXAMPLE 16 Multiply $(3x - 4)$ by $(2x + 5)$.

$$
\begin{array}{l}
3x \;\; - \;\; 4 \\
2x \;\; + \;\; 5 \\
\hline
6x^2 - \;\; 8x \\
\;\;\;\; + 15x - 20 \\
\hline
6x^2 + \;\; 7x - 20
\end{array}
$$

EXAMPLE 17 Multiply $(3x - 2y)$ by $(3x - y)$.

$$
\begin{array}{l}
3x \;\; - 2y \\
3x \;\; - \;\; y \\
\hline
9x^2 - 6xy \\
\;\;\;\; - 3xy + 2y^2 \\
\hline
9x^2 - 9xy + 2y^2
\end{array}
$$

This method is good for multiplying longer polynomials as follows.

EXAMPLE 18 Multiply $(x + 2)$ by $(x^2 + 2x - 5)$.

$$
\begin{array}{l}
x^2 + 2x \;\; - 5 \\
\;\;\;\;\; x \;\; + 2 \\
\hline
x^3 + 2x^2 - 5x \qquad \text{This is } x(x^2 + 2x - 5) \\
\;\;\;\; + 2x^2 + 4x - 10 \qquad \text{This is } 2(x^2 + 2x - 5) \\
\hline
x^3 + 4x^2 - \;\; x - 10
\end{array}
$$

EXAMPLE 19   Multiply $(a - b)$ by $(a^2 + ab + b^2)$.

$$
\begin{array}{l}
a^2 + ab \ + b^2 \\
\underline{\qquad a \ - b} \\
a^3 + a^2b + ab^2 \\
\underline{\quad - a^2b - ab^2 - b^3} \\
a^3 + 0 \quad + 0 \quad - b^3
\end{array}
$$

which is $a^3 - b^3$.

EXAMPLE 20   What does $(x + 2)(x^2 + 2x - 5)$ stand for when x is 3?

We will do this example in two ways.

Method 1:

We put in 3 for x.

$(x + 2)(x^2 + 2x - 5)$

$(3 + 2) \, (3^2 + 2(3) - 5)$

$(5)(9 + 6 - 5)$

$(5)(10)$ which is 50

Method 2:

From example 18 we know that $(x + 2)(x^2 + 2x - 5)$ is the same as $x^3 + 4x^2 - x - 10$. So let's put 3 in for x in that expression.

$x^3 + 4x^2 - x - 10$

$3^3 + 4(3)^2 - 3 - 10$

$27 + 4(9) - 3 - 10$

$27 + 36 - 3 - 10$ which is 50

Exercises

What single number does each of the following stand for?
1. $3(4 + 5)$
2. $4(-2 + 1)$
3. $5(3 - 2)$
4. $3(4 + 2 + 1)$
5. $2(6 + 3 + 2)$
6. $5(7 - 4 - 1)$

Multiply
7. $2(x + y)$
8. $3(a - b)$
9. $4(a + b + c)$
10. $5(x + 1)$
11. $3(2x + 5y)$
12. $5(3a + 7b)$
13. $4(x + 3y - 5z)$
14. $3x(2 + 7x)$
15. $4x(3x^2 + 8)$
16. $a(x - y)$
17. $5x(x - y + z)$
18. $-3(x - 4y)$
19. $(2 + 7x)3x$
20. $(3x - 5y)8x$
21. $(5x - 2)x^2$
22. $(x^2 + y^2)xy$
23. $(a^2 - b^2)a^2b^2$
24. $(x + y)z + (x + y)a$
25. $(x + y)z + z(x + y)$
26. $3(x + y) + (4 + 5x)y$
27. $ab(2a - 3b + 5)$
28. $x^2y(x - 5y + 1)$
29. $5xy(4x - 1 + 7xy)$
30. $4ab^2(2a + 3b - 4)$
31. $-3xyz(5x - 3yz + y)$
32. $-4abc(2a - 3b^2 + 5c)$
33. $-5x^2y^3z^4(4x^2 - 2 + y^2)$
34. $(x + 3)(x + 4)$
35. $(x + 5)(x + 1)$

36. $(x - 3)(x - 2)$
37. $(x + 6)(x - 3)$
38. $(2y + 3)(y + 4)$
39. $(3a - 2)(2a - 5)$
40. $(4b + 5)(b - 2)$
41. $(5c - 6)(3c + 2)$
42. $(x^2 + 3x + 2)(x + 4)$
43. $(x^2 + 4x - 3)(x - 5)$
44. $(a^2 + 5a - 6)(a - 4)$
45. $(2x^2 + 5x - 8)(x - 2)$
46. $(a^2 - ab + b^2)(a + b)$
47. $(2x - 5y)^2$
48. $(x - 1)(x + 7)(x + 2)$

Answers to Exercises

1. $3(4+5)$ or $3(4+5)$
   $3(9)$ $12+15$
   27 27

2. $4(-2+1)$ or $4(-2+1)$
   $4(-1)$ $-8+4$
   -4 -4

3. $5(3-2)$
   $15-10$
   5

4. $3(4+2+1)$
   $12+6+3$
   21

5. $2(6+3+2)$
   $12+6+4$
   22

6. $5(7-4-1)$
   $35-20-5$
   10

7. $2(x+y)$
   $2x+2y$

8. $3(a-b)$
   $3a-3b$

9. $4(a+b+c)$
   $4a+4b+4c$

10. $5(x+1)$
    $5x+5$

11. $3(2x+5y)$
    $6x+15y$

12. $5(3a+7b)$
    $15a+35b$

13. $4(x+3y-5z)$
    $4x+12y-20z$

14. $3x(2+7x)$
    $6x+21x^2$

15. $4x(3x^2+8)$
    $12x^3+32x$

16. $a(x-y)$
    $ax-ay$

17. $5x(x-y+z)$
$5x^2 - 5xy + 5xz$

18. $-3(x-4y)$
$-3x + 12y$

19. $(2+7x)3x$
$6x + 21x^2$

20. $(3x-5y)8x$
$24x^2 - 40xy$

21. $(5x \cdot 2)x^2$
$5x^3 - 2x^2$

22. $(x^2 + y^2)xy$
$x^3y + xy^3$

23. $(a^2 - b^2)a^2b^2$
$a^4b^2 - a^2b^4$

24. $(x+y)z + (x+y)a$
$xz + yz + ax + ay$

25. $(x+y)z + z(x+y)$
$xz + yz + xz + yz$
$2xz + 2yz$

26. $3(x+y) + (4+5x)y$
$3x + 3y + 4y + 5xy$
$3x + 7y + 5xy$

27. $ab(2a - 3b + 5)$
$2a^2b - 3ab^2 + 5ab$

28. $x^2y(x-5y+1)$
$x^3y - 5x^2y^2 + x^2y$

29. $5xy(4x - 1 + 7xy)$
$20x^2y - 5xy + 35x^2y^2$

30. $4ab^2(2a + 3b - 4)$
$8a^2b^2 + 12ab^3 - 16ab^2$

31. $-3xyz(5x - 3yz + y)$
$-15x^2yz + 9xy^2z^2 - 3xy^2z$

32. $-4abc(2a - 3b^3 + 5c)$
$-8a^2bc + 12ab^3c - 20abc^2$

33. $-5x^2y^3z^4(4x^2-2+y^2)$

$-20x^4y^3z^4+10x^2y^3z^4-5x^2y^5z^4$

34. $(x+3)(x+4)$
$(x+3)x+(x+3)4$
$x^2+3x+4x+12$
$x^2+7x+12$

35. $x+5$
$\underline{x+1}$
x^2+5x
$\underline{1x+5}$
x^2+6x+5

36. $x-3$
$\underline{x-2}$
x^2-3x
$\underline{-2x+6}$
x^2-5x+6

37. $x+6$
$\underline{x-3}$
x^2+6x
$\underline{-3x-18}$
$x^2+3x-18$

38. $2y+3$
$\underline{y+4}$
$2y^2+3y$
$\underline{+8y+12}$
$2y^2+11y+12$

39. $3a-2$
$\underline{2a-5}$
$6a^2-4a$
$\underline{-15a+10}$
$6a^2-19a+10$

40. $(4b+5)(b-2)$
$(4b+5)b+(4b+5)(-2)$
$4b^2+5b-8b-10$
$4b^2-3b-10$

41. $5c-6$
$\underline{3c+2}$
$15c^2-18c$
$\underline{+10c-12}$
$15c^2-8c-12$

㊷ $(x^2+3x+2)(x+4)$

$(x^2+3x+2)x + (x^2+3x+2)4$

$x^3+3x^2+2x+4x^2+12x+8$

$x^3+7x^2+14x+8$

㊸ x^2+4x-3
　　　　$x-5$
x^3+4x^2-3x
　　$-5x^2-20x+15$
$x^3-x^2-23x+15$

㊹ a^2+5a-6
　　　$a-4$
a^3+5a^2-6a
　　$-4a^2-20a+24$
$a^3+a^2-26a+24$

㊺ $2x^2+5x-8$
　　　　$x-2$
$2x^3+5x^2-8x$
　　$-4x^2-10x+16$
$2x^3+x^2-18x+16$

㊻ a^2-ab+b^2
　　　$a+b$
$a^3-a^2b+ab^2$
　　$+a^2b-ab^2+b^3$
$a^3\qquad\qquad+b^3$

㊼ $(2x-5y)(2x-5y)$
　　$2x-5y$
　　$2x-5y$
$4x^2-10xy$
　　$-10xy+25y^2$
$4x^2-20xy+25y^2$

㊽ $(x-1)(x+7)(x+2)$
　　$x-1$
　　$x+7$
x^2-1x
　　$+7x-7$
x^2+6x-7 then

x^2+6x-7
　　$x+2$
x^3+6x^2-7x
　　$2x^2+12x-14$
$x^3+8x^2+5x-14$

Additional Exercises

What single number does each of the following stand for?

1. $2(3 + 7)$
2. $5(-3 + 2)$
3. $4(6 - 1)$
4. $2(5 + 3 + 1)$
5. $3(7 + 4 + 2)$
6. $5(6 - 3 - 1)$

Multiply the following.

7. $3(a + b)$
8. $4(x - y)$
9. $5(a - b + c)$
10. $7(a - 1)$
11. $4(3x + 7y)$
12. $6(2a + 5b)$
13. $5(3x - 5y + 7)$
14. $2x(3 + 8x)$
15. $5x(6x^2 - 7)$
16. $b(y - z)$
17. $7(a - b + c)$
18. $-3(a + b)$
19. $-3(a - b)$
20. $-4(x - 2y)$
21. $(3 + 9x)4x$
22. $(5x - 6y)2x$
23. $(3a - 5)a^2$
24. $(a^2 - b^2)ab$
25. $(x^2 - y^2)x^3y^3$
26. $(a + b)x + (a + b)c$
27. $(b + c)d + d(b + c)$
28. $4(x + 2y) + (5 - 3x)y$
29. $xy(3x - 5y + 2)$
30. $a^2b^2(4a - 7b + 3)$
31. $2xy(6xy - 7x + 1)$
32. $3a^2b(5a - 4b + 6)$
33. $-2abc(3a + 5b + 9)$
34. $-7xyz(3xyz + 4x^4 - 5y)$
35. $-6x^3y^2z(5 - 2x + 4xyz)$
36. $(x + 5)(x + 2)$
37. $(x + 7)(x + 3)$
38. $(x - 4)(x - 8)$
39. $(x + 9)(x - 1)$

40. $(3a + 4)(a + 2)$
41. $(2a - 5)(3a - 4)$
42. $(5y + 6)(y - 4)$
43. $(6b - 5)(2b + 3)$
44. $(x^2 + 4x + 3)(x + 2)$
45. $(a^2 + 6a - 4)(a - 3)$
46. $(3x^2 + 4x - 6)(x - 5)$
47. $(5x^2 + 2x - 7)(2x - 3)$
48. $(x^2 - 3xy + y^2)(x - y)$
49. $(3x - 4y)^2$
50. $(x - 2)(x - 4)(x + 1)$

Lesson 7

Factoring

In Lesson 6, we found that $2(a + b)$ is the same as $2a + 2b$. We did this by multiplying everything inside the parentheses by 2. If we start with $2a + 2b$ and we write it as $2(a + b)$, we call this **factoring**. We have **factored out** the 2.

$$2a + 2b \quad \text{is factored as} \quad 2(a + b)$$

It is important to know how to go back and forth between these two forms. Sometimes it is convenient to have an expression as a <u>sum</u>, like $2a + 2b$, and sometimes it is convenient to have an expression as a <u>product</u>, like $2(a + b)$.

Let's look at $3x + 6y$.

$$3x + 6y \quad \text{is the same as} \quad (3)(x) + (2)(3)(y)$$

Note that we have split the first term into two parts (factors): 3 and x. We have split the second term into three parts (factors): 2 and 3 and y. Since 3 is a factor of $3x$ and since 3 is a factor of $6y$, 3 is called a **common factor** of $3x$ and $6y$.

We can factor out the common factor, 3, as follows.

$$3x + 6y$$

is the same as $(3)(x) + (2)(3)(y)$

$$3(x + 2y)$$

We can check our factoring by multiplying everything inside the parentheses by 3. We should get back the original expression.

$$3(x + 2y) \quad \text{is} \quad 3x + 6y \qquad \text{It checks!}$$

Let's write $6x^2 + 4x$ as a product by factoring out all the common factors.

$$6x^2 + 4x$$

is the same as $(2)(3)(x)(x) + (2)(2)(x)$

2 and x are common factors. Let's factor them out.

$$(2)(x)(3x + 2)$$

is $2x(3x + 2)$

In this problem, we call $2x$ the **complete common factor**, since $2x$ contains all the common factors.

We have factored $6x^2 + 4x$ as $2x(3x + 2)$. We can check by multiplying $2x$ by $(3x + 2)$.

$$2x(3x + 2) \text{ is } 6x^2 + 4x$$

So $6x^2 + 4x$ can be factored as $2x(3x + 2)$.

EXAMPLE 1   Write $10ab + 5abc + 15bc$ as a product by factoring out all the common factors. That is, factor the expression completely.

$$10ab + 5abc + 15bc$$

is the same as $(2)(5)(a)(b) + (5)(a)(b)(c) + (3)(5)(b)(c)$

Now look for a factor common to each term. Since 5 appears in each term, it is a common factor. Let's underline each 5.

$$(2)\underline{(5)}(a)(b) + \underline{(5)}(a)(b)(c) + (3)\underline{(5)}(b)(c)$$

But b also appears in each term, so it is also a common factor. Let's underline each b as well.

$$(2)\underline{(5)}(a)\underline{(b)} + \underline{(5)}(a)\underline{(b)}(c) + (3)\underline{(5)}\underline{(b)}(c)$$

There are no other common factors, so $5b$ is the complete common factor. Now we factor out $5b$.

$$10ab + 5abc + 15bc \quad \text{becomes} \quad 5b(2a + ac + 3c)$$

Note: What is underlined comes out of the parentheses; what is not underlined is left behind.

So $10ab + 5abc + 15bc$ can be written as the product

$5b(2a + ac + 3c)$.

EXAMPLE 2   Factor $x^3y - 3x^2y^2$ completely.

$$x^3y - 3x^2y^2$$

can be written $xxxy - 3xxyy$

As before, we can underline the common factors.

$$\underline{xxxy} - 3\underline{xxyy}$$

The only common factors are the x and x and y; therefore, the complete common factor is xxy, and the factored form is

$xxy(x - 3y)$

or $x^2y(x - 3y)$

To check, we multiply x^2y by $(x - 3y)$ to get $x^3y - 3x^2y^2$. Thus $x^3y - 3x^2y^2$ can be factored as $x^2y(x - 3y)$.

EXAMPLE 3   Factor $2 + 2b$ completely.

We can underline the common factor 2.

$\underline{2} + \underline{2}b$.

Since the entire first term is underlined, the problem is what to put inside the parentheses when we factor out the 2.

$2 + 2b$ is the same as $2(? + b)$

So $2 + 2b$ is the same as $2(?) + 2b$

Therefore 2 must be the same as $2(?)$ and the ? must be 1. We have factored $2 + 2b$ as $2(1 + b)$.

We can look at the problem in a different way.

$2 + 2b$ is $(2)(1) + (2)(b)$

We can underline the common factor 2.

$\underline{(2)}(1) + \underline{(2)}(b)$

So the factored form is $2 + 2b$ is $2(1 + b)$.

EXAMPLE 4 Factor $2x^2 - 6xy - 2x$ completely.

$$2x^2 - 6xy - 2x$$

can be written as $(2)(x)(x) - (2)(3)(x)(y) - (2)(x)$

Underline the common factors as follows.

$\underline{(2)}\underline{(x)}(x) - \underline{(2)}(3)\underline{(x)}(y) - \underline{(2)}\underline{(x)}$

$2x$ is the complete common factor.

Note: Each factor in the last term is underlined.

So let's rewrite the last term as $-(2)(x)(1)$.

$\underline{(2)}\underline{(x)}(x) - \underline{(2)}(3)\underline{(x)}(y) - \underline{(2)}\underline{(x)}(1)$

can be factored as $2x(x - 3y - 1)$

Let's check it by multiplying $2x$ by $x - 3y - 1$ to get $2x^2 - 6xy - 2x$.

So the factored form of $2x^2 - 6xy - 2x$ is $2x(x - 3y - 1)$.

EXAMPLE 5   Factor $2ac + 3ab - 6bc$ completely.

This can be written as $(2)(a)(c) + (3)(a)(b) - (2)(3)(b)(c)$.

<u>Dead End!</u>   There are no common factors, this polynomial cannot be factored. Expressions that can't be factored are called **prime**.

EXAMPLE 6   Factor $-5x + 10y$ completely.

$$-5x + 10y$$

can be written as $-(5)(x) + (5)(2)(y)$

Underline the common factors as follows.

$$-\underline{(5)}(x) + \underline{(5)}(2)(y)$$

So $-5x + 10y$ can be factored as $5(-x + 2y)$.

It is sometimes more convenient to factor out -5 rather than 5. If we write 5 as $(-5)(-1)$, then -5 is a common factor as follows.

$$\underline{(-5)}(x) + \underline{(-5)}(-1)(2)(y)$$ *factoring out − by using*

Thus $-5x + 10y$ can also be factored as $-5(x - 2y)$. Let's check both factored forms.

$5(-x + 2y)$ is the same as $-5x + 10y$

$-5(x - 2y)$ is the same as $-5x + 10y$

So $-5x + 10y$ can be factored as $5(-x + 2y)$ or $-5(x - 2y)$.

EXAMPLE 7   Factor $12x^2yz^2 - 6x^3y^2z + 9x^4yz^3$ completely.

This polynomial can be written as

$$(3)(2)(2)xxyzz - (3)(2)xxxyyz + (3)(3)xxxxyzzz$$

$(3)(x)(x)(y)(z)$ is the complete common factor. Therefore the polynomial can be factored as

$$3x^2yz(4x - 2xy + 3x^2z^2)$$

EXAMPLE 8   Factor $a^{25}b^{35}c^{40} + a^{23}b^{30}c^{42}$ completely.

Let's take a shortcut. Look at the two terms and find a useful way to split up each one.

$$a^{25}b^{35}c^{40} + a^{23}b^{30}c^{42}$$

$$(\underline{a^{23}})(\underline{a^2})(\underline{b^{30}})(\underline{b^5})(\underline{c^{40}}) + (\underline{a^{23}})(\underline{b^{30}})(\underline{c^{40}})(c^2)$$

The polynomial

$$a^{25}b^{35}c^{40} + a^{23}b^{30}c^{42}$$

can thus be factored as

$$a^{23}b^{30}c^{40}(a^2b^5 + c^2)$$

Check it to be sure.

EXAMPLE 9   What does $a^2b - 5ab + ab^2$ stand for when a is 2 and b is 3?

We will do this example two ways.

Method 1:

We put in 2 for a and 3 for b.

$a^2b - 5ab + ab^2$

$(2^2)(3) - 5(2)(3) + 2(3^2)$

$4(3) - 5(6) + 2(9)$

$12 - 30 + 18$ which is   0

Method 2:

We will first factor the expression.

$a^2b - 5ab + ab^2$ can be factored as $ab(a - 5 + b)$

Now we put in 2 for a and 3 for b

$ab(a - 5 + b)$

$2(3)(2 - 5 + 3)$

$6(0)$ which is 0

So by either method, when a is 2 and b is 3, the expression $a^2b - 5ab + ab^2$ stands for 0.

Exercises

Write each of the following expressions as a product by factoring out all the common factors. That is, factor completely. Be careful—some of these can't be factored; expressions that can't be factored are called **prime**.

1. $3x + 3y$
2. $5a - 5b$
3. $2x + 4$
4. $xy + xz$
5. $ab + ac - ad$
6. $a^2 - a + ad$
7. $ax^2 + a$
8. $a^2b + a^2c + a^2d$
9. $a^2b + a^3c - a^4d$
10. $3x^2 - 6xy + ax$
11. $6xy + 3x$
12. $2x - 6x^2$
13. $6a^2 + 4ab - 10ac$
14. $10a^2b - 5a$
15. $2x^2y^2 + xy^3$
16. $9x + 3xy$
17. $9x^3y^2 - 6x^4y$
18. $6x^2y^3 + 3xy^4$
19. $x^2 - 2x^2y$
20. $5x^2 + 10x^3y^3$
21. $10x^2 + 15x^2y - 5x$
22. $3x^2y + 7x + 5y^2z$
23. $3xy + 2x^2z - 6yz^3$
24. $18x^2yz^3 - 9xy^3 + 3x^4y^2z$
25. $6a^3bc^4 - 4a^2b^2c^3 + 2a^2bc^4$
26. $11x^{54}y^{12}z^6 - 22x^3y^5z^8 + 33x^{17}y^4z^5$
27. $3x^2yz^4 + x^2y - 3z^3$
28. $13x^2yz^3 + 7x^3$
29. $3x^{26}y^6z^9 - 2x^{25}y^5z^{12}$
30. $3a^{15}b^7c^{23} - 27a^{15}$

Do the following word problems.

31. What does $x^2y + xy^2$ stand for when x is 3 and y is 2?

32. What does $x^2y + xy^2$ stand for when x is 1 and y is -3?

33. What does $a^2b^2 - 4ab$ stand for when a is 2 and b is 3?

34. What does $a^2b^2 - 4ab$ stand for when a is 4 and b is -2?

Answers to Exercises

① $3x + 3y$
  $\underline{3}x + \underline{3}y$
  $3(x+y)$

② $5a - 5b$
  $\underline{5}a - \underline{5}b$
  $5(a-b)$

③ $2x + 4$
  $\underline{2}x + \underline{2} \cdot 2$
  $2(x+2)$

④ $xy + xz$
  $\underline{x}y + \underline{x}z$
  $x(y+z)$

⑤ $ab + ac - ad$
  $\underline{a}b + \underline{a}c - \underline{a}d$
  $a(b+c-d)$

⑥ $a^2 - a + ad$
  $\underline{a}a - 1\cdot\underline{a} + \underline{a}d$
  $a(a-1+d)$

⑦ $ax^2 + a$
  $\underline{a}xx + \underline{a}$
  $a(x^2+1)$

⑧ $a^2b + a^2c + a^2d$
  $\underline{a\,a}b + \underline{a\,a}c + \underline{a\,a}d$
  $a^2(b+c+d)$

⑨ $a^2b + a^3c - a^4d$
  $\underline{a\,a}b + \underline{a\,a}ac - \underline{a\,a}aad$
  $a^2(b+ac-a^2d)$

⑩ $3x^2 - 6xy + ax$
  $3\underline{x}x - 2\cdot3\,\underline{x}y + a\,\underline{x}$
  $x(3x-6y+a)$

⑪ $6xy + 3x$
  $2\cdot\underline{3}\,\underline{x}y + 1\cdot\underline{3}\underline{x}$
  $3x(2y+1)$

⑫ $2x - 6x^2$
  $1\cdot\underline{2}\underline{x} - \underline{2}\cdot3\underline{x}x$
  $2x(1-3x)$

⑬ $6a^2 + 4ab - 10ac$
  $\underline{2}\cdot3\underline{a}a + \underline{2}\cdot2\underline{a}b - \underline{2}\cdot5\underline{a}c$
  $2a(3a+2b-5c)$

⑭ $10a^2b - 5a$
  $2\cdot\underline{5}\underline{a}ab - 1\cdot\underline{5}\underline{a}$
  $5a(2ab-1)$

(15) $2x^2y^2 + xy^3$
$2xxyy + xyyy$
$xy^2(2x+y)$

(16) $9x + 3xy$
$3\cdot3x + 3xy$
$3x(3+y)$

(17) $9x^3y^2 - 6x^4y$
$3\cdot3xxxyy - 2\cdot3xxxxy$
$3x^3y(3y - 2x)$

(18) $6x^2y^3 + 3xy^4$
$2\cdot3xxyyy + 3xyyyy$
$3xy^3(2x+y)$

(19) $x^2 - 2x^2y$
$1\cdot x\cdot x - 2xxy$
$x^2(1-2y)$

(20) $5x^2 + 10x^3y^3$
$1\cdot5xx + 2\cdot5xxxyyy$
$5x^2(1+2xy^3)$

(21) $10x^2 + 15x^2y - 5x$
$2\cdot5xx + 3\cdot5xxy - 5x$
$5x(2x + 3xy - 1)$

(22) $3x^2y + 7x - 5y^2z$
$3xxy + 7x - 5gyz$
no common factors
It is a prime!

(23) $3xy + 2x^2z - 6yz^3$
$3xy + 2xxz - 2\cdot3yzzz$
prime!

(24) $18x^2yz^3 - 9xy^3 + 3x^4y^2z$
$3xy(6xz^3 - 3y^2 + x^3yz)$

(25) $6a^3bc^4 - 4a^2b^2c^3 + 2a^2bc^4$
$2a^2bc^3(3ac - 2b + c)$

(26) $11x^{54}y^{12}z^6 - 22x^3y^5z^8 + 33x^{17}y^4z^5$
$11x^3y^4z^5(x^{51}y^8z - 2yz^3 + 3x^{14})$

27. $3x^2yz^4 - x^2y + 3z^3$
$3xxy\ zzzz - xxy + 3zzz$
no common factors!
prime!

28. $13x^2yz^3 + 7x^3$
$13\underline{xx}y\ zzz + 7\underline{xxx}$
$x^2(13yz^3 + 7x)$

29. $3x^{26}y^6z^9 - 2x^{25}y^5z^{12}$
$3\underline{x^{25}}xy^5yz^9 - 2\underline{x^{25}}y^5\underline{z^9}z^3$
$x^{25}y^5z^9(3xy - 2z^3)$

30. $3a^{15}b^7c^{23} - 27a^{15}$
$\underline{3}a^{15}b^7c^{23} - \underline{3}\cdot9\ \underline{a^{15}}$
$3a^{15}(b^7c^{23} - 9)$

31. $x^2y + xy^2$
$xy(x+y)$
$\underline{(3)(2)}\ (3+2)$
$(6)\quad (5)$
30

32. $x^2y + xy^2$
$xy(x+y)$
$\underline{(1)(-3)}\ (1-3)$
$(-3)\quad (-2)$
6

33. $a^2b^2 - 4ab$
$ab(ab-4)$
$(2)(3)\ ((2)(3)-4)$
$6\ (6-4)$
$6(2)$
12

34. $a^2b^2 - 4ab$
$ab(ab-4)$
$(4)(-2)\ ((4)(-2)-4)$
$(-8)\ (-8-4)$
$(-8)(-12)$
96

Additional Exercises

Write each of the following expressions as a product by factoring out all the common factors. That is, factor completely. Be careful—some of these can't be factored; expressions that can't be factored are called **prime**.

1. $2a + 2b$
2. $6x - 6y$
3. $8a + 8b - 8c$
4. $3x + 9$
5. $ab - ac$
6. $xy + xz - xw$
7. $bc^2 + b$
8. $3x + 6y - 9z$
9. $2y^2 - 8xy + 4y$
10. $ax^2 + bx^3 + cx^4$
11. $3a + 2ab$
12. $xy - x$
13. $5xy^2 - xy$
14. $16x^2y + 12xy^2$
15. $3a^2b + 2ab^2$
16. $9x^5y^2z - 3xy^2z^2$
17. $15a^3bc^2 + 5a^2c$
18. $7x^5y^2z^3 - 14x^2yz^5$
19. $3x^2 + 5y^3$
20. $3a^2b + 2ac^2 - 7bc^2$
21. $3x^7y^3z^5 + 6x^4y^4z^4 - 9x^3y^7z^2$
22. $15x^{40}y^{30}z^{20} - 5x^{39}y^{29}z^{19}$
23. $12x^5y^3z^5 + 6xz^3 - 18x^4y^2z^6$
24. $5a^9b^3c^4 + 10a^7b^9c^7 - 15a^4b^3c$
25. $3ab^2 + 9a^2c - 6bc^5$
26. $17x^2b + 7xb^3$
27. $7x^2y + 3xz - 2x^3y$
28. $x^{37}y^3 - x^{39}y^5$
29. $x^7y^8z^5 + x^9y^5z^4 - x^2y^5z^3$
30. $a^2b^3c - a^9b^4c^3 + a^4b^7c^6$

Do the following word problems.

31. What does $x^3y + xy^3$ stand for when x is 2 and y is -4?

32. What does $x^3y + xy^3$ stand for when x is -2 and y is 5?

33. What does $b^2c^2 - 5bc$ stand for when b is 3 and c is 2?

34. What does $b^2c^2 - 5bc$ stand for when b is -3 and c is 4?

Lesson 8

Division of Polynomials by Monomials

In this lesson we will divide polynomials by monomials.

$\dfrac{6 + 8}{2}$ is an example of a polynomial (6 + 8) divided by a monomial (2).

Here are two ways of dealing with this division problem:

$$\dfrac{6 + 8}{2} \quad \text{is} \quad \dfrac{14}{2} \quad \text{which is} \quad 7$$

and also $\quad \dfrac{6 + 8}{2} \quad$ is $\quad \dfrac{6}{2} + \dfrac{8}{2} \quad$ which is $\quad 3 + 4 \quad$ or $\quad 7$

Similarly $\quad \dfrac{6a + 8b}{2}$ can be written as $\quad \dfrac{6a}{2} + \dfrac{8b}{2}$.

So when you have more than one term (a polynomial) on top and only one term (a monomial) on the bottom, you can break the whole thing into several fractions, each one with the same bottom.

118

EXAMPLE 1   Divide $6a + 8b$ by 2.

$$\frac{6a + 8b}{2}$$

is the same as $\dfrac{6a}{2} + \dfrac{8b}{2}$

which is $3a + 4b$

So $\dfrac{6a + 8b}{2}$ is $3a + 4b$.

EXAMPLE 2   Divide $1.5r + 4.5s + 5$ by 0.5.

$$\frac{1.5r + 4.5s + 5}{0.5}$$

is the same as $\dfrac{1.5r}{0.5} + \dfrac{4.5s}{0.5} + \dfrac{5}{0.5}$

which is $3r + 9s + 10.$

EXAMPLE 3

$$\frac{3a^2 + 5a + 6ab}{a}$$

is the same as $\dfrac{3a^2}{a} + \dfrac{5a}{a} + \dfrac{6ab}{a}$

which is $\dfrac{3aa}{a} + \dfrac{5a}{a} + \dfrac{6ab}{a}$

or $3a + 5 + 6b$

This example can be done in a different way. In Lesson 7 we factored polynomials. The top of the above fraction is a polynomial and can be factored as follows.

$3a^2 + 5a + 6ab$ can be written as $a(3a + 5 + 6b)$

So $\dfrac{3a^2 + 5a + 6ab}{a}$ can be written as $\dfrac{a(3a + 5 + 6b)}{a}$

Note: a is a factor on the top and the bottom.

We can cancel and replace $\dfrac{a}{a}$ with 1.

So $\dfrac{a(3a + 5 + 6b)}{a}$ is the same as $\dfrac{\not{a}(3a + 5 + 6b)}{\not{a}}$

and $\dfrac{3a^2 + 5a + 6ab}{a} = 3a + 5 + 6b.$

EXAMPLE 4   Simplify $\dfrac{6x^2y - 3xy}{3xy}$.

$$\frac{6x^2y - 3xy}{3xy}$$

is the same as $\dfrac{6x^2y}{3xy} - \dfrac{3xy}{3xy}$

which is $\dfrac{2(3)x\not{x}\not{y}}{3\not{x}\not{y}} - \dfrac{3xy}{3xy}$

which is $\qquad 2x - 1$

Note: $\dfrac{3xy}{3xy}$ is 1. Don't lose it!

This example also can be done by factoring and canceling. Since $6x^2y - 3xy$ can be factored as $3xy(2x - 1)$, we can write

$$\frac{6x^2y - 3xy}{3xy}$$

$3xy(2x-1)$

as $\dfrac{\not{3}\not{x}\not{y}(2x - 1)}{\not{3}\not{x}\not{y}}$

or $\quad 2x - 1$

So $\dfrac{6x^2y - 3xy}{3xy} = 2x - 1.$

This kind of problem can be done either way. Choose the one you like.

EXAMPLE 5   Divide $2bc^2 - 4b^2c^2 + 6bc^3$ by $2bc^2$.

$$\frac{2bc^2 - 4b^2c^2 + 6bc^3}{2bc^2}$$

is the same as $\dfrac{2bc^2}{2bc^2} - \dfrac{4b^2c^2}{2bc^2} + \dfrac{6bc^3}{2bc^2}$

which is $\dfrac{2bc^2}{2bc^2} - \dfrac{2(2)bbc^2}{2bc^2} + \dfrac{2(3)bc^2c}{2bc^2}$

which is $1 - 2b + 3c$

So $\dfrac{2bc^2 - 4b^2c^2 + 6bc^3}{2bc^2} = 1 - 2b + 3c.$

EXAMPLE 6   Simplify $\dfrac{2f^2y - 4fy + 6fy^3}{2fy}$.

The top of the fraction can be factored, so

$\dfrac{2f^2y - 4fy + 6fy^3}{2fy}$ is the same as $\dfrac{2fy(f - 2 + 3y^2)}{2fy}$

Now we can cancel.

$\dfrac{\cancel{2fy}(f - 2 + 3y^2)}{\cancel{2fy}}$

So $\dfrac{2f^2y - 4fy + 6fy^3}{2fy}$ is the same as $f - 2 + 3y^2.$

EXAMPLE 7   Simplify $\dfrac{-10c^2d^2 + 5c^3d^4 + 15c^4d^5}{-5c^2d^2}$.

$\dfrac{-10c^2d^2 + 5c^3d^4 + 15c^4d^5}{-5c^2d^2}$

is the same as $\dfrac{-10c^2d^2}{-5c^2d^2} + \dfrac{5c^3d^4}{-5c^2d^2} + \dfrac{15c^4d^5}{-5c^2d^2}$

which is $\dfrac{-5(2)c^2d^2}{-5c^2d^2} + \dfrac{5c^2cd^2d^2}{-5c^2d^2} + \dfrac{5(3)c^2c^2d^2d^3}{-5c^2d^2}$

which is $2 - cd^2 - 3c^2d^3.$

EXAMPLE 8   Simplify $\dfrac{12x^5y^7z^{15} - 18x^4y^6z^{11} - 6x^7y^9z^{10}}{6x^4y^6z^{10}}$.

We factor the top of the fraction and then cancel.

$\dfrac{\cancel{6x^4y^6z^{10}}(2xyz^5 - 3z - x^3y^3)}{\cancel{6x^4y^6z^{10}}}$

is $2xyz^5 - 3z - x^3y^3.$

EXAMPLE 9 Simplify $\dfrac{2(3a + 4ab) - 3(2a^3 + 6ab^2)}{2a}$.

This expression contains parentheses, so we first multiply to get rid of them.

$$\frac{2(3a + 4ab) - 3(2a^3 + 6ab^2)}{2a}$$

is the same as $\dfrac{6a + 8ab - 6a^3 - 18ab^2}{2a}$

which is $\dfrac{6a}{2a} + \dfrac{8ab}{2a} - \dfrac{6a^3}{2a} - \dfrac{18ab^2}{2a}$

or $3 + 4b - 3a^2 - 9b^2$

There are no like terms, so that's all we can do.

EXAMPLE 10 Evaluate $\dfrac{5a^2b + 15ab^2}{5ab}$ when a is 2.5 and b is -1.7.

We could put in 2.5 for a and -1.7 for b and then do the calculations. However, it is easier to first simplify the expression.

$\dfrac{5a^2b + 15ab^2}{5ab}$ becomes $\dfrac{\cancel{5ab}(a + 3b)}{\cancel{5ab}}$

Now we put in 2.5 for a and -1.7 for b.

$a + 3b$

$2.5 + 3(-1.7)$

$2.5 - 5.1$

which is -2.6.

Exercises

1. Divide $(10 + 4)$ by 7
2. Divide $(15 + 50)$ by 5
3. Divide $(50 - 30)$ by 10
4. Divide $(8 - 12)$ by -2
5. Divide $(-2x + 6y)$ by 2
6. Divide $(3x - 15y)$ by 3
7. Divide $(x^2 + 2x)$ by x
8. Divide $(-3xy + 6yz)$ by $-3y$
9. Divide $(x^2 + xy)$ by x
10. Divide $(x^3 - x^2)$ by x^2
11. Divide $(x^2y + xy + xy^2)$ by $-xy$
12. Divide $(3x + xy - x^2)$ by x
13. Divide $(4x^2y^3 - 8xy^3)$ by $4xy^3$
14. Divide $(27a^2bc - 9ab^2c + 18abc^2)$ by $-9abc$

Do each of the following division problems.

15. $\dfrac{2x^2 + 4xy}{2x}$

16. $\dfrac{16x^7y^{12} - 10x^5y^8}{2x^2y^7}$

17. $\dfrac{a^3b^2c - a^2bc^3 - ab^3c}{abc}$

18. $\dfrac{4a^2b - 2ab}{2ab}$

19. $\dfrac{3xy^2 - 6x^2y^2 + 9xy^3}{3xy^2}$

20. $\dfrac{4c^2d - 8c^2d^2 + 12cd^2}{4cd}$

21. $\dfrac{14x^6y^7z^{12} - 21x^4y^5z^{10} - 7x^5y^7z^{15}}{7x^4y^5z^{10}}$

22. $\dfrac{24a^9b^{12}c^{15} - 18a^7b^7c^{10} + 12a^5b^{10}c^{10}}{6a^4b^5c^{10}}$

23. Evaluate $\dfrac{4x^2y + 12xy^2}{4xy}$ when x is 3 and y is 2.

24. Evaluate $\dfrac{3ab + 6a^2b^2}{3ab}$ when a is -2 and b is -3.

25. Evaluate $\dfrac{8b^2c - 12bc^2}{4bc}$ when b is 1.5 and c is 2.4.

26. Evaluate $\dfrac{2x^3y^3 - 6x^2y^3}{2x^2y^2}$ when x is 3.4 and y is -1.5.

Answers to Exercises

(1.) $\dfrac{10+4}{7}$

$= \dfrac{14}{7}$

$= 2$

(2.) $\dfrac{15+50}{5}$ or $\dfrac{15+50}{5}$

$= \dfrac{65}{5}$ $\quad = \dfrac{15}{5} + \dfrac{50}{5}$

$= 13$ $\quad\quad = 3 + 10$

$\quad\quad\quad\quad = 13$

(3.) $\dfrac{50-30}{10}$

$= \dfrac{50}{10} - \dfrac{30}{10}$

$= 5 - 3$

$= 2$

(4.) $\dfrac{8-12}{-2}$

$= \dfrac{8}{-2} - \dfrac{12}{-2}$

$= -4 + 6$

$= +2$

(5.) $\dfrac{-2x+6y}{2}$

$= \dfrac{-2x}{2} + \dfrac{6y}{2}$

$= -x + 3y$

(6.) $\dfrac{3x-15y}{3}$

$= \dfrac{3x}{3} - \dfrac{15y}{3}$

$= x - 5y$

(7.) $\dfrac{x^2+2x}{x}$

$= \dfrac{x^2}{x} + \dfrac{2x}{x}$

$= x + 2$

(8.) $\dfrac{-3xy+6yz}{-3y}$

$\dfrac{-3xy}{-3y} + \dfrac{6yz}{-3y}$

$x - 2z$

(9.) $\dfrac{x^2+xy}{x}$

$\dfrac{x^2}{x} + \dfrac{xy}{x}$

$x + y$

(10.) $\dfrac{x^3 - x^2}{x^2}$

$\dfrac{x^3}{x^2} \quad -\dfrac{1x^2}{x^2}$

$x - 1$

(12.) $\dfrac{3x + xy - x^2}{x}$

$\dfrac{3x}{x} + \dfrac{xy}{x} - \dfrac{xx}{x}$

$3 + y - x$

(11.) $\dfrac{x^2 y + xy + xy^2}{-xy}$

$\dfrac{xxy}{-xy} + \dfrac{xy}{-xy} + \dfrac{xyy}{-xy}$

$-x - 1 - y$

(13.) $\dfrac{4x^2 y^3 - 8x\, y^3}{4x\, y^3}$

$\dfrac{xxxyyy}{xxyyy} - \dfrac{2 \cdot xx\, yyy}{xxyyy}$

$x - 2$

(14.) $\dfrac{27a^2 bc - 9a\, b^2 c + 18abc^2}{-9abc}$

$\dfrac{3 \cdot 9a\, abb}{-9abbc} - \dfrac{9abbc}{-9abbc} + \dfrac{2 \cdot 9a\, bbc}{-9abb}$

$-3a + b - 2c$

(15.) $\dfrac{2x^2 + 4xy}{2x}$

$\dfrac{2xx}{2x} + \dfrac{xxy}{2x}$

$x + 2y$

(16.) $\dfrac{16x^7 y^{12} - 10x^5 y^8}{2x^2 y^7}$

$\dfrac{16x^7 y^{12}}{2x^2 y^7} - \dfrac{10x^5 y^8}{2x^2 y^7}$

$8x^5 y^5 - 5x^3 y$

(17.) $\dfrac{a^3 b^2 c}{abc} - \dfrac{a^2 bc^3}{abc} - \dfrac{ab^3 c}{abc}$

$a^2 b - ac^2 - b^2$

(18.) $\dfrac{4a^2b - 2ab}{2ab}$

$\dfrac{\cancel{2ab}(2a-1)}{\cancel{2ab}}$

$2a - 1$

(19.) $\dfrac{3xy^2 - 6x^2y^2 + 9xy^3}{3xy^2}$

$\dfrac{\cancel{3xy^2}(1 - 2x + 3y)}{\cancel{3xy^2}}$

$1 - 2x + 3y$

(20.) $\dfrac{4c^2d - 8c^2d^2 + 12cd^2}{4cd}$

$\dfrac{\cancel{4cd}(c - 2cd + 3d)}{\cancel{4cd}}$

$c - 2cd + 3d$

(21.) $\dfrac{14x^6y^7z^{12} - 21x^4y^5z^{10} - 7x^5y^7z^{15}}{7x^4y^5z^{10}}$

$\dfrac{\cancel{7x^4y^5z^{10}}(2x^2y^2z^2 - 3 - xy^2z^5)}{\cancel{7x^4y^5z^{10}}}$

$2x^2y^2z^2 - 3 - xy^2z^5$

(22.) $\dfrac{24a^9b^{12}c^{15} - 18a^7b^7c^{10} + 12a^5b^{10}c^{10}}{6a^4b^5c^{10}}$

$\dfrac{\cancel{6a^5b^7c^{10}}(4a^4b^5c^5 - 3a^2 + 2b^3)}{\cancel{6a^4b^5c^{10}}}$

$ab^2(4a^4b^5c^5 - 3a^2 + 2b^3)$

(23.) $\dfrac{4x^2y + 12xy^2}{4xy}$

$\dfrac{\cancel{4xy}(x + 3y)}{\cancel{4xy}}$

$x + 3y$

$3 + 3(2)$
$3 + 6$
9

(24.) $\dfrac{3ab + 6a^2b^2}{3ab}$

$\dfrac{\cancel{3ab}(1 + 2ab)}{\cancel{3ab}}$

$1 + 2ab$
$1 + 2(-2)(-3)$
$1 + 12$
13

25.) $\dfrac{8b^2c - 12\,bc^2}{4bc}$

$\dfrac{\cancel{4bc}\,(2b - 3c)}{\cancel{4bc}}$

$2b - 3c$

$2(1.5) - 3(2.4)$

$3 - 7.2$

-4.2

26.) $\dfrac{2x^3y^3 - 6x^2y^3}{2x^2y^2}$

$\dfrac{\cancel{2}\cancel{x^2}y^3\,(x-2)}{\cancel{2}\cancel{x^2}y^2}$

$y\,(x-2)$

$(-1.5)(3.4-2)$

$(-1.5)(1.4)$

-2.1

why is factoring appropriate for 24 and not 20?

Additional Exercises

1. Divide $(16 - 4)$ by 3
2. Divide $(26 + 2)$ by -7
3. Divide $(100 - 25)$ by 25
4. Divide $(50 - 20)$ by -6
5. Divide $(x + xy)$ by x
6. Divide $(4a - 12b)$ by 4
7. Divide $(-x^4 + x^2)$ by x
8. Divide $(x^3 - 5x)$ by x
9. Divide $(x^2y^3 - xy)$ by xy
10. Divide $(x^5 - x^4)$ by x^2
11. Divide $(-3x^2y - 6xy^3)$ by $3xy$
12. Divide $(4a^3 + 6a^2)$ by $-2a^2$
13. Divide $(ab + ac - ad)$ by a
14. Divide $(16b^3 - 24b^5 - 8b^2)$ by $8b^2$

Do each of the following division problems.

15. $\dfrac{3a^2 + 6ab}{3a}$

16. $\dfrac{16a^4b^2 - 8a^2b^3}{4a^2b^2}$

17. $\dfrac{x^3y^2z - x^2y^2z + xyz^3}{xyz}$

18. $\dfrac{6b^2c - 3bc^2}{3bc}$

19. $\dfrac{5x^2y - 10x^2y^2 + 15xy^2}{5xy}$

20. $\dfrac{3ab - 12a^3b^2 + 9ab^2}{3ab}$

21. $\dfrac{24x^8y^9z^{12} - 32x^9y^9z^9 + 16x^{15}y^{14}z^{11}}{8x^5y^7z^9}$

22. $\dfrac{-12a^8b^{11}c^{13} + 8a^7b^7c^8 + 4a^9b^6c^8}{4a^7b^6c^8}$

23. Evaluate $\dfrac{5a^2b + 10a^2b^2}{5ab}$ when a is 4 and b is 3.

24. Evaluate $\dfrac{4xy + 8x^2y^2}{4xy}$ when x is -3 and y is -2.

25. Evaluate $\dfrac{9r^2t - 18rt^2}{3rt}$ when r is 2.2 and t is 1.4.

26. Evaluate $\dfrac{3x^4y^4 - 6x^3y^4}{3x^3y^3}$ when x is 4.1 and y is -2.5.

Index